Lecture Notes in Artificial Intelligence   5465

Edited by R. Goebel, J. Siekmann, and W. Wahlster

Subseries of Lecture Notes in Computer Science

Debbie Richards   Byeong-Ho Kang (Eds.)

Knowledge Acquisition: Approaches, Algorithms and Applications

Pacific Rim Knowledge Acquisition Workshop, PKAW 2008
Hanoi, Vietnam, December 15-16, 2008
Revised Selected Papers

 Springer

Series Editors

Randy Goebel, University of Alberta, Edmonton, Canada
Jörg Siekmann, University of Saarland, Saarbrücken, Germany
Wolfgang Wahlster, DFKI and University of Saarland, Saarbrücken, Germany

Volume Editors

Debbie Richards
Macquarie University, Computing Department
Division of Information and Communication Sciences
Sydney, NSW, 2109, Australia
E-mail: richards@ics.mq.edu.au

Byeong-Ho Kang
University of Tasmania
School of Computing and Information Systems
Launceston, TAS 7250, Australia
E-mail: bhkang@utas.edu.au

Library of Congress Control Number: Applied for

CR Subject Classification (1998): I.2.6, I.2, H.2.8, H.3-5, F.2.2, C.2.4, K.3

LNCS Sublibrary: SL 7 – Artificial Intelligence

ISSN   0302-9743
ISBN-10   3-642-01714-2 Springer Berlin Heidelberg New York
ISBN-13   978-3-642-01714-8 Springer Berlin Heidelberg New York

springer.com

© Springer-Verlag Berlin Heidelberg 2009
Printed in Germany

Typesetting: Camera-ready by author, data conversion by Scientific Publishing Services, Chennai, India
Printed on acid-free paper   SPIN: 12676811   06/3180   5 4 3 2 1 0

Preface

With the growing recognition of the pivotal role that knowledge plays in the sustainability, competitiveness and growth of individuals, organizations and society, finding solutions to address the knowledge acquisition bottleneck is even more important today than in the early stages of this field. The knowledge acquisition community is interested in topics spanning from the fundamental views on knowledge that affect the knowledge acquisition process and the use of knowledge in knowledge engineering to the evaluation of knowledge acquisition techniques, tools and methods.

As a field within the larger field of artificial intelligence (AI), solutions incorporating other areas of AI such as ontological engineering, agent-based technology, robotics, image recognition and the Semantic Web are common, as are knowledge acquisition methods related to other fields of computing such as databases, the Internet, information retrieval, language technology, software engineering, decision support systems and game technology. Many solutions are application focused addressing real-world problems such as knowledge maintenance and validation, reuse and sharing, merging and reconciliation within a wide range of problem domains.

The Pacific Knowledge Acquisition Workshops (PKAW) have provided a forum for the past two decades for researchers and practitioners in the Pacific region and beyond who work in the field of knowledge acquisition. PKAW covers a spectrum of techniques and approaches ranging from manual knowledge acquisition from a human expert to fully automated knowledge acquisition using machine-learning or data-mining methods.

This volume seeks to disseminate the latest solutions from the Pacific Knowledge Acquisition Workshop 2008 (PKAW 2008) held in Hanoi, Taiwan during December 15-16, 2008 in conjunction with the Pacific Rim International Conference on Artificial Intelligence (PRICAI 2008). The workshop received 57 submissions from 14 countries. From these, we accepted 15 papers (26%) for full presentation and another 14 for short presentation. All papers were blind reviewed by at least three members of the Program Committee. This volume contains a selection of these papers further revised following workshop discussions. The papers demonstrate a balance of theoretical, technical and application-driven research, many papers incorporating all three foci. Approximately half the papers reflect the increasing use of Web-based data for knowledge discovery and management.

The Workshop Co-chairs would like to thank all those who were involved in PKAW 2008 including the PRICAI 2008 Organizing Committee, PKAW Program Committee members, those who submitted papers and reviewed them and

of course the authors, presenters and attendees. We warmly invite you to participate in PKAW 2010 anticipated to be held in Seoul, Korea in conjunction with
PRICAI 2010.

March 2009   Debbie Richards
                                                    Byeong Ho Kang

Organization

Honorary Chairs

Paul Compton   University of New South Wales, Australia
Hiroshi Motoda   Osaka University and AFOSR/AOARD, Japan

Program Co-Chairs

Debbie Richards   Macquarie University, Australia
Byeong Ho Kang   University of Tasmania, Australia

Program Committee

Ghassan Beydoun	University of Wollongong, Australia
Bob Colomb	University of Technology Malaysia, Malaysia
Paul Compton	University of New South Wales, Australia
Richard Dazeley	University of Ballarat, Australia
Fabrice Guillet	LINA - Polytech'Nantes, France
Udo Hahn	Jena University, Germany
Byeong Ho Kang	University of Tasmania, Australia
Rob Kremer	University of Calgary, Canada
Maria Lee	Shih Chien University, Taiwan
Tim Menzies	LCSEE, WVU, USA
Toshiro Minami	Kyushu University, Japan
Hiroshi Motoda	Osaka University and AFOSR/AOARD, Japan
Kozo Ohara	Osaka University, Japan
Takashi Okada	Kwansei Gakuin University, Japan
Son Pham	College of Technology, VNU, Vietnam
Frank Puppe	University of Würzburg, Germany
Ulrich Reimer	University of Applied Sciences St. Gallen, Switzerland
Debbie Richards	Macquarie University, Australia
Hendra Suryanto	University of New South Wales, Australia
Seiji Yamada	National Institute of Informatics, Japan
Kenichi Yoshida	University of Tsukuba, Japan
Tetsuya Yoshida	Hokkaido Univerisity, Japan

Additional Reviewers

Daniel Bidulock
Jason Heard
Uwe Heck
Claudia Marinica
Christian Thiel
Ryan Yee

Table of Contents

Domain Specific Knowledge Acquisition Methods and Applications

Experiments with Adaptive Transfer Rate in Reinforcement Learning

Yann Chevaleyre[1], Aydano Machado Pamponet[2], and Jean-Daniel Zucker[3]

[1] Universit Paris-Dauphine
yann.chevaleyre@lamsade.dauphine.fr
[2] Universit Paris 6
aydano.machado@lip6.fr
[3] IRD UR Godes
Centre IRD de l'Ile de France, Bondy, France
jean-daniel.zucker@ird.fr

Abstract. Transfer algorithms allow the use of knowledge previously learned on related tasks to speed-up learning of the current task. Recently, many complex reinforcement learning problems have been successfully solved by efficient transfer learners. However, most of these algorithms suffer from a severe flaw: they are implicitly tuned to transfer knowledge between tasks having a given degree of similarity. In other words, if the previous task is very dissimilar (resp. nearly identical) to the current task, then the transfer process might slow down the learning (resp. might be far from optimal speed-up). In this paper, we address this specific issue by explicitly optimizing the transfer rate between tasks and answer to the question : "can the transfer rate be accurately optimized, and at what cost ?". We show that this optimization problem is related to the continuum bandit problem. We then propose a generic adaptive transfer method (AdaTran), which allows to extend several existing transfer learning algorithms to optimize the transfer rate. Finally, we run several experiments validating our approach.

1   Introduction

In the reinforcement learning problem, an agent acts in an unknown environment, with the goal of maximizing its reward. All learning agents have to face the exploration-exploitation dilemma: whether to act so as to explore unknown areas or to act consistently with experience to maximize reward (exploit). Most research on reinforcement learning deal with this issue. Recently Strehl et al. [17] showed that near optimal strategies could be reached in as few as $\widetilde{O}(S \times A)$ time steps. However, in many real-world learning problems, the state space or the action space have an exponential size.

One way to circumvent this problem is to use previously acquired knowledge related to the current task being learned. This knowledge may then be used to guide exploration through the state-action space, hopefully leading the agent towards areas in which high rewards can be found. This knowledge can be acquired in different ways:

D. Richards and B.-H. Kang (Eds.): PKAW 2008, LNAI 5465, pp. 1–11, 2009.

- By imitation: in particular, in a multi-agent environment, agents may observe the behavior of one another and use this observation to improve their own strategy [13].
- By bootstrap: related tasks may have been previously learned by reinforcement [14,1,10] and the learned policy may be used to bootstrap the learning.
- By abstraction: a simplified version of the current task could have been generated to quickly learn a policy which could be used as a starting point for the current task [18,21].
- By demonstration: a human tutor may provide some explicit knowledge. Other similar settings exist in the literature, among which "advice taking" [19] or "apprenticeship" [20].

In this paper, we will focus on a simple version of the "bootstrap" transfer learning problem [1]: we will assume that a policy (or a Q-value function) is available to the learner, and that this policy has been learned on a past task which shares the *same state-action space* as that of the current task.

Given this knowledge, the learning agent faces a new dilemma: it has to balance among following the ongoing learned policy and exploring the available policy. Most transfer learners do not tackle this dilemma explicitly: the amount of exploration based on the available policy does not depend on its quality. Ideally, this amount should be tuned such that the transfert learner be *robust* w.r.t. the quality of the past policy : good policies should speedup the learner while bad ones should not slow it down significantly. Recently, a new approach has been proposed to solve this issue [10,11]. The main idea of this approach is to estimate such similarity between the two tasks, and then to use this estimate as a parameter of the transfer learning process, balancing between ongoing and past policies. However, measuring this similarity is a costly process in itself and moreover there are no guarantee that this similarity optimizes the transfer learning process.

In this paper, we show that a parameter called the *transfer rate* controling the balance between past policy and the ongoing policy can be optimized efficiently *during* the reinforcement learning process. For this purpose, we first show in which way this optimization problem is related to the *continuum-armed bandit* problem. Based on this relation, we propose a generic adaptive transfer learning method (AdaTran) consisting of a wrapper around some standard transfer learning algorithm, and implementing a continuum-armed bandit algorithm.

We choose two representative transfer approaches, namely *Probabilistic Policy Reuse* (PPR, [1]) and *Memory-guided Exploration* (MGE, [12]) which, wrapped inside AdaTran, will be referred to as AdaTran(PPR) and AdaTran(MGE). We show experimentally on a grid-world task that AdaTran is much more robust than non adaptive transfer algorithms.

The paper is organized as follows. After some preliminaries and a state of the art on transfer learners, we introduce the continuous bandit problem and relate it to the optimization of the transfer rate. The following section introduces the AdaTran framework and its instantitation AdaTran(PPR) and AdaTran(MGE). Then, a set of experiments in which AdaTran is compared to standard transfer learners assesses both its robustness and efficiency.

2 Preliminaries

Reinforcement learning problems are typically formalized using Markov Decision Processes (MDPs). An MDP M is a tuple $\langle S, A, T, r, \gamma \rangle$ where S is the set of all states, A is the set of all actions, T is a state transition function $T : S \times A \times S \rightarrow \mathbb{R}$, r is a reward function $r : S \times A \rightarrow \mathbb{R}$, and $0 \leq \gamma < 1$ is a discount factor on rewards. From a state s under action a, the agent receives a stochastic reward r, which has expectation $r(s, a)$, and is transported to state s' with probability $T(s, a, s')$. A policy is a strategy for choosing actions. If it is also deterministic, a policy can be represented by a function $\pi : S \rightarrow A$. As in most transfer learning settings, we assume that the learning process is divided into episodes : at the beginning of an episode, the agent is placed on a starting state sampled from a distribution $\mathcal{D}$. The episode ends when the agent reaches a special absorbing state (the goal), or when a time limit is reached.

For any policy π, let $V_M^\pi(s)$ denote the discounted value function for π in M from state s. More formally, $V_M^\pi(s) \triangleq \mathbb{E}\left[\sum_{t=0}^\infty \gamma^t r_t\right]$, where $r_0, r_1, \ldots$ is the reward sequence obtained by following policy π from state s. Also, let $V_M^\pi \triangleq \mathbb{E}_{s \sim \mathcal{D}}[V_M^\pi(s)]$. To evaluate the quality of an action under a given policy, the Q-value function $Q^\pi(s, a) \triangleq r(s, a) + \gamma \mathbb{E}_{s' \sim T(s,a,.)}\left[V^\pi(s')\right]$ is generally used (Here, as there are no ambiguity, M has been omitted). The optimal policy π^* is the policy maximizing the value function. The goal of any reinforcement learning algorithm is to find a policy such that the agent's performance approaches that of π^*.

To speed up learning on a new task, transfer learners exploit knowledge previously learned on a past task. Here, we will assume an in [1] that the past task and the current task have the same state-action space. We study the case where the available knowledge has the form of a policy $\bar{\pi}$ learned on the past task or of a Q-value function $\bar{Q}$. Both cases lead to different transfer learners, such as PPR and MGE.

3 Transfer Learners with Static Transfer Rates

In this section, we will present two state-of-the-art transfer learners, namely PPR (*Probabilistic Policy Reuse*, as well as PPR-decay, a variation on PPR [1]) and MGE (Memory-Guided Exploration [12]), exhibiting a parameter which controls the balance between the ongoing learned policy and $\bar{\pi}$. As most transfer methods, PPR and MGE have been directly build on a standard Q-learner, and thus share the same structure. The only difference with a Q-learner lies in the action selection method (referred here as *ChooseAction*).

The most widely used transfer method probably is Q-reuse ([14], also sometimes referred to as *direct transfer*). Based on a Q-learner, this method simply uses $\bar{Q}$ to initialize the Q-values of the current task. Caroll *et al* note in [12] that with this approach, "if the tasks are too dissimilar, the agent will spend too much time unlearning...". To overcome this drawback, they propose an alternative to Q-reuse, namely *Memory-Guided Exploration*, a standard Q-learner in which the action selection procedure has been replaced as shown in table 1. Here, ξ is a

Table 1. Examples of $ChooseAction(s_t, \bar{\pi}, \varphi)$ functions in static transfer learners

$ChooseAction(s_t, \bar{\pi}, \varphi)$	Name of transfer algorithm
$a_t = argmax_a a \left\{ (1 - \varphi)Q_t(s_t, a) + \varphi Q_t(s_t, a) + \xi \right\}$ where ξ is some real-valued random variable.	MGE [12] (Memory Guided Exploration)
$a_t = \begin{cases} \bar{\pi}(s_t) & w.\, proba. \varphi \times 0,95^t \\ \epsilon\text{-greedy}(\pi) & otherwise \end{cases}$	PPR-decay [1] (PPR with exponential decay)
$a_t = \begin{cases} \bar{\pi}(s_t) & with\, proba.\, \varphi \\ \epsilon\text{-greedy}(\pi) & with\, proba\, 1 - \varphi \end{cases}$	PPR (Probabilistic Policy Reuse)

random variable (e.g. normal distributed with zero mean) used to add random exploration. Clearly, the parameter φ influences the procedure: if $\varphi \to 0$, then the algorithm is similar to a standard Q-learner, and if $\varphi \to 1$, it always follow $\bar{\pi}$.

Recently, Fernandez and Veloso proposed a completely different probabilistic approach to transfer learning (PPR-decay [1]) also based on a Q-learner. At each step, the algorithm randomly chooses to follow the policy ϵ-greedy(π) or to follow $\bar{\pi}$, as depicted in table 1. Here, π refers to the policy induces by the Q-values $(\pi(s) = argmax_a Q_t(s, a))$ and ϵ-greedy(π) refers to the policy obtained by choosing π with probability $1 - \epsilon$, or a random action with probability ϵ. Fernandez et al. proposed arbitrarily to initialize φ to one at the beginning of each episode, and to decrease its influence at each step t by $0,95^t$. As for MGE, PPR-decay mimics a Q-learner when $\varphi = 0$, but does not follow $\bar{\pi}$ at each step when $\varphi = 1$, because of the decay. Therefor, we introduce a variation on PPR-decay, namely PPR, in which φ is not decreased during the episode.

Clearly, φ can be seen here as a parameter controlling the *transfer rate*, although it does not have exactly the same role in PPR and MGE. It is not hard to see that this rate should be dependent on the similarity between the past and the current task. Computing such a similarity is difficult in the general case, and that optimizing φ can be done during learning, as shown in the next sections.

4   Optimization of the Transfer Rate as a Stochastic Continuum-Armed Bandit Problem

Consider a transfer method such as one of those discussed above, in which a parameter $\varphi \in [0, 1]$ controls the transfer rate, in such a way that if $\varphi = 0$, the policy $\bar{\pi}$ is not being used, and if $\varphi = 1$, the agent follows exclusively $\bar{\pi}$. Let us consider the problem of optimizing φ, in order to improve the speedup learning. For the sake of simplicity, adjustment of φ will occur only after each episode, thus exploiting the sequence of rewards gathered during the last episode.

Consider a learning episode starting at time t. Before the episode begins, the agent has to choose a value of φ, which ideally would yield the highest expected gain $V_t(\varphi) \triangleq \mathbb{E}\left[r_t + \gamma r_{t+1} + \dots \mid \varphi\right]$. At the end of the episode, the agent can compute $\sum_k r_{t+k}\gamma^k$ which is an unbiased estimator of $V_t(\varphi)$. Choosing the best value for φ is challenging, as gradient methods which require the knowledge of

$\frac{\partial V_t}{\partial \varphi}$ might not be applicable. It turns out that this problem is a typical *multi-armed bandit problem*.

The *continuum armed-bandit* problem which belongs to the well known family of multi-armed bandit problems, is a particularly appropriate setting for the optimization of $V_t(\varphi)$. In [8], a stochastic version of this problem is presented, for which the specific algorithm UCBC was designed. In [7], an algorithm called CAB1 is described, in order to solve the *adversarial* version of this problem. In our setting, the gain $\sum_k r_{t+k}\gamma^k$ is stochastic, and changing in time. Thus, we will need a generalization of these two settings, namely the *stochastic adversarial continuum armed-bandit problem*, which can be described as follows

Definition 1. (The stochastic adversarial continuum armed-bandit problem) *Assume the existence of an unknown distribution family $P(. \mid x,t)$ indexed by $x \in [0,1]$ and $t \in \{1\ldots n\}$. At each trial t, the learner chooses $X_t \in [0,1]$ and receives return $Y_t \sim P(. \mid X_t, t)$. Let $b_t(x) \triangleq \mathbb{E}[Y_t \mid X_t = x, t]$. The agent's goal is to minimize its expected regret $\mathbb{E}[\sum_t b_t(x^*) - \sum_t Y_t]$, given that $x^* = \sup_{x \in [0,1]} \sum_{t=1}^{n} b_t(x)$.*

Although this setting seems more general than the adversarial case, the regret guarantees of CAB1 still hold here. Investigating regret bounds is beyond the scope of this paper. However, the reader can refer to theorem 3.1 of [7], which proof can easily be generalized to the stochastic adversarial setting.

5   AdaTran: A Generic Adaptive Transfer Framework

We now present a generic adaptive transfer learning algorithm, which can be seen as a wrapper around a transfer learner, optimizing the transfer rate φ using a *stochastic adversarial continuum armed-bandit* algorithm referred to as *UpdateContBandit*. This leads to the *AdaTran* wrapper, a generic adaptive transfer algorithm in which many transfer learners can be implemented. Note that even though most transfer learners do not have such a parameter, they can often be modified so as to make φ appear explicitly.

Algorithm 1. AdaTran

1: Init()
2: $t \leftarrow 0$
3: $\varphi \leftarrow \varphi_0$
4: **for** each episode h **do**
5:     set the initial state s
6:     **while** (end of episode not reached) **do**
7:         $a_t = ChooseAction(s_t, \bar{\pi}, \varphi)$
8:         Take action a_t, observe r_{t+1}, s_{t+1}
9:         $Learn(s_t, a_t, r_{t+1}, s_{t+1})$
10:        $t \leftarrow t + 1$
11:    **end while**
12:    $\varphi \leftarrow UpdateContBandit(\bar{\pi}, \varphi, \langle r_1, r_2, \ldots \rangle)$
13: **end for**

Depending of the function used for *ChooseAction* (e.g. one of table 1), *Learn* (e.g. a TD update of a model-based learning step) and *UpdateContBandit* (e.g. CAB1), the AdaTran will lead to different types of transfer learners. In particular, the experimental section will evaluate AdaTran(PPR), AdaTran(PPR-decay), and AdaTran(MGE).

Let us now show how CAB1 can be applied. Let t_i be the time at which the ith episode begins. Let φ_t refer to the parameter chosen by CAB1 at time t. Then on the n first episodes, CAB1 will try to minimize the regret $\sum_{i=1}^{n} V_{t_i}(\varphi^*) - V_{t_I}(\varphi_t)$, which consists in finding the best transfer rate.

6   Experiments

In this section, we evaluate AdaTran on a standard benchmark for transfer learning: the grid-world problem [1,6]. The reason behind our choice of this learning task lies in its simplicity. As the state space is discrete, and no function approximation method is required, experimental results are more likely to evaluate the transfer approaches per se instead and not so much the whole machinery needed to make a complex learning problem tractable. In this learning task, an agent moves in a 25×25 two-dimensional maze. Each cell of this grid-world is a state and it may be surrounded by zero to four walls. Each cell has two coordinates (x, y), and the cell at the center of the grid has coordinates $(0, 0)$. At each time step, the agent can choose to move from its current position to one of the reachable contiguous north/east/west/south cell. If a wall lies in between, the action fails. Otherwise, the move succeeds with probability 90%, and with probability 10%, the agent is randomly placed on one of the reachable cells contiguous to the current cell. At the beginning of each episode, the agent is randomly and uniformly placed on the maze. As the agent reaches the goal state, it is given a reward of 1, and the episode is ended. All other rewards are null and the discount factor is arbitrarily set to $\gamma = 0, 95$. At the beginning of each episode, the learning agent is place at the center of the grid, at position $(0, 0)$. The goal of the current task T is to reach the bottom right corner at $(12, -12)$. We generated three tasks T_1, T_2 and T_3 having three different goal positions $(12,-10),(12,12),(-12,12)$.

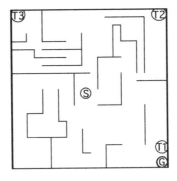

Fig. 1. grid-world

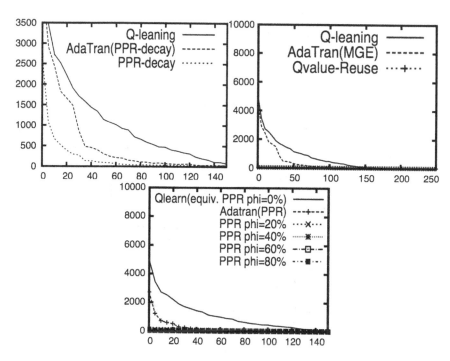

Fig. 2.   Evaluation   of   AdaTran(PPR-decay),   AdaTran(PPR-decay)   and AdaTran(PPR-decay) against other learners, using T_1 as the transfer task

The optimal policies computed on each of these three tasks will serve as transfer knowledge to solve T. The goals of T_1 and T are very close to each other. Thus, transfer between both might be highly valuable. On the opposite, the goals in T_3 and T are very dissimilar to one another, and transfer is likely to be less valuable. In between, T and T_2 can be seen as "orthogonal" to each other: moving towards the goal of T_2 does not make the goal of T closer or farther. This will allow us to evaluate the robustness of AdaTran compared to other algorithms. Each of the following curves have been averaged over 100 runs. The x-axis represents the episodes, and the y-axis is the average episode length, given that episode are limited to 10000 steps.

Let us first consider the experiments with T_1 as a transfer task (fig. 2). As expected, the Q-learner performs worst, unable to exploit a task very similar to the current task. Also as expected, Qvalue-Reuse and "PPR phi" with $\varphi \geq$ 20% perform extremely well, as they are biased to follow the transfer policy often. Even though the performance of AdaTran on this task are not as good as that of these biased transfer learners, compared to the Q-learner, AdaTran performs much better. For example, after 40 episodes, the average duration of a Q-learning episode is three times that of an AdaTran(PPR), AdaTran(PPR-decay) or AdaTran(MGE) episode.

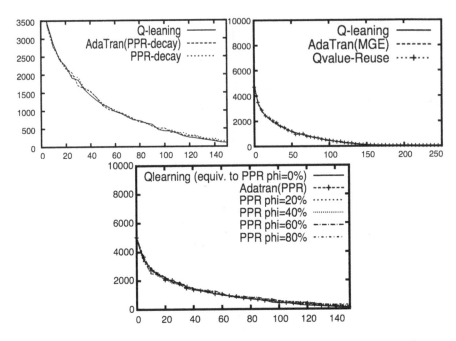

Fig. 3. Evaluation of AdaTran(PPR-decay), AdaTran(PPR-decay) and AdaTran(PPR-decay) against other learners, using T_2 as the transfer task

Concerning the transfer between T_2 and T, the "orthogonality" of both tasks appears in the curves (fig. 3). No transfer method achieves better or worse results than Q-learning.

Concerning the transfer from T_3 to T, which are very dissimilar tasks, we note that Q-learning always achieve best results, as it is never mislead by the transfer policy. Also note that AdaTran is quite close to Q-learning, which is the best it could do : the bandit in AdaTran must learns as fast as possible not to follow the transfer policy. It can be seen on figure 4 that PPR-decay is the best of the non-adaptive transfer learners here. In fact, PPR-decay follows the transfer policy only at the beginning of each episode, and is thus less penalized than others by bad transfer policies. Also, the Qvalue-reuse algorithm, which is one of the most used transfer method in reinforcement learning, performs very badly : as seen on figure 4, it requires approximately 150 episodes to "unlearn" the transfer policy. On the contrary, AdaTran(MGE) which also reuses the Q-values selects low values of φ very quickly, thus not using the transfer knowledge much. Lastly, note that PPR-phi are the worst non-adaptive transfer learners, as the episode length (y-axis) always reaches it maximal value. Clearly, AdaTran is shown to be robust to dissimilar tasks (T_3) unlike the other transfer methods studied here, and transfer successfully a high amount of knowledge on similar tasks (T_1).

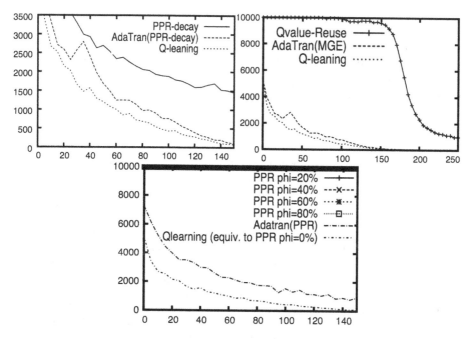

Fig. 4. Evaluation of AdaTran(PPR-decay), AdaTran(PPR-decay) and AdaTran(PPR-decay) against other learners, using T_3 as the transfer task

7   Conclusion and Future Work

In this paper, we have presented a new framework for explicitly optimizing the transfer rate in reinforcement learning. We have shown how this framework could be applied on two representative transfer learners to make the transfer rate auto-adaptable, namely the *probabilistic policy reuse* methods and *MGE* method, related to the well known Qvalue-reuse.

We have shown experimentally that AdaTran is robust to misleading transfer knowledge: when the transfer task is similar to the current task, AdaTran's performance will be close to non-adaptive transfer methods, but when the transfer task is very dissimilar to the current task, AdaTran will not spend a large amount of time forgetting transfer knowledge, unlike non-adaptive transfer learners.

There are four main directions for future work. First, the performance of these algorithms must be evaluated on real-world learning tasks. Some preliminary experiments on various grid-world tasks suggest that AdaTran makes transfer learners much more robust but it remains to be precisely characterized. Another important issue concerns the optimization criterion. In this paper, we proposed to optimize $V(\varphi)$. However, this implies trying to maximize the short term gain, which is heavily biased towards exploitation. Because of the close relation between our approach and gradient policy search methods, we are studying the criteria used in the latter [2].

Also, theoretical work relating the regret of the bandits to the expected performance of AdaTran remains to be done. Finally, we could design bandit algorithms taking the specificity of our problem into account. In fact, the adversarial setting may be excessive for our application, and the function $V(\varphi)$ probably has some properties (such as some kind of monotonicity through time) which may be exploited favorably.

References

1. Fernandez, F., Veloso, M.: Probabilistic Policy Reuse in a Reinforcement Learning Agent. In: The Fifth International Joint Conference on Autonomous Agents and Multiagent Systems, AAMAS (2006)
2. Kakade, S.: On The Sample Complexity of Reinforcement Learning. Phd Thesis, University College London (2003)
3. Taylor, M.E., Stone, P.: Behavior Transfer for Value-Function-Based Reinforcement Learning. In: The Fourth International Joint Conference on Autonomous Agents and Multiagent Systems, July 2005, pp. 53–59. ACM Press, New York (2005)
4. Taylor, M.E., Whiteson, S., Stone, P.: Transfer via Inter-Task Mappings in Policy Search Reinforcement Learning. In: The Sixth International Joint Conference on Autonomous Agents and Multiagent Systems, AAMAS 2007 (May 2007)
5. Perkins, T.J., Precup, D.: Using options for knowledge transfer in reinforcement learning. Technical report, University of Massachusetts, Amherst (1999)
6. Price, B., Boutilier, C.: Accelerating Reinforcement Learning through Implicit Imitation. Journal of Artificial Intelligence Research 19, 569–629 (2003)
7. Kleinberg, R.: Nearly tight bounds for the continuum-armed bandit problem. In: Advances in Neural Information Processing Systems 17 (NIPS 2004), pp. 697–704 (2004)
8. Auer, P., Ortner, R., Szepesvri, C.: Improved Rates for the Stochastic Continuum-Armed Bandit Problem. In: Bshouty, N.H., Gentile, C. (eds.) COLT. LNCS, vol. 4539, pp. 454–468. Springer, Heidelberg (2007)
9. Auer, P., Cesa-Bianchi, N., Freund, Y., Schapire, R.E.: Gambling in a Rigged Casino: The adversarial multi-armed bandit problem. In: Proceedings of the 36th Annual Symposium on Foundations of Computer Science (1998)
10. Carroll, J.L., Peterson, T.S.: Fixed vs. Dynamic Sub-transfer in Reinforcement Learning. In: Arif Wani, M. (ed.) ICMLA 2002, Las Vegas Nevada, USA, June 24-27, 2002. CSREA Press (2002)
11. Carroll, J.L., Seppi, K.: Task Similarity Measures for Transfer in Reinforcement Learning Task Libraries. In: The 2005 International Joint Conference on Neural Networks, IJCNN 2005 (2005)
12. Carroll, J.L., Peterson, T.S., Owens, N.E.: Memory-guided Exploration in Reinforcement Learning. In: The 2001 International Joint Conference on Neural Networks, IJCNN 2001 (2001)
13. Price, B., Boutilier, C.: A Bayesian Approach to Imitation in Reinforcement Learning. In: IJCAI 2003, pp. 712–720 (2003)
14. Taylor, M.E., Stone, P., Liu, Y.: Transfer Learning via Inter-Task Mappings for Temporal Difference Learning. Journal of Machine Learning Research 8(1), 2125–2167 (2007)
15. Zhou, D.-X.: A note on derivatives of Bernstein polynomials. Journal of Approximation Theory 78(1), 147–150 (1994)
16. Lorentz, G.G.: Bernstein Polynomials. Chelsea, New York (1986)

17. Strehl, A.L., Li, L., Wiewiora, E., Langford, J., Littman, M.L.: PAC model-free reinforcement learning. In: ICML 2006: Proceedings of the 23rd international conference on Machine learning, pp. 881–888 (2006)
18. Madden, M.G., Howley, T.: Transfer of Experience Between Reinforcement Learning Environments with Progressive Difficulty. Artif. Intell. Rev. 21(3-4), 375–398 (2004)
19. Torrey, L., Walker, T., Shavlik, J., Maclin, R.: Knowledge Transfer Via Advice Taking. In: Proceedings of the Third International Conference on Knowledge Capture, KCAP 2005 (2005)
20. Abbeel, P., Ng, A.Y.: Apprenticeship learning via inverse reinforcement learning. In: Proceedings of the Twenty-first International Conference on Machine Learning (2004)
21. Konidaris, G.D.: Autonomous Shaping: Knowledge Transfer in Reinforcement Learning. In: Proceedings of the Twenty Third International Conference on Machine Learning (ICML 2006), Pittsburgh, PA (June 2006)
22. Kearns, M., Singh, S.: Finite-Sample Rates of Convergence for Q-Learning and Indirect Methods. In: Advances in Neural Information Processing Systems 11, pp. 996–1002. The MIT Press, Cambridge (1999)

Clustering over Evolving Data Streams Based on Online Recent-Biased Approximation

Wei Fan[1], Yusuke Koyanagi[1], Koichi Asakura[2], and Toyohide Watanabe[1]

[1] Department of Systems and Social Informatics,
Graduate School of Information Science, Nagoya University
Furo-cho, Chikusa-ku, Nagoya, 464-8603, Japan
[2] School of Informatics, Daido Institute of Technology
10-3 Takiharu-cho, Minami-ku, Nagoya, 457-8530, Japan
fan@watanabe.ss.is.nagoya-u.ac.jp

Abstract. A growing number of real world applications deal with multiple evolving data streams. In this paper, a framework for clustering over evolving data streams is proposed taking advantage of recent-biased approximation. In recent-biased approximation, more details are preserved for recent data and fewer coefficients are kept for the whole data stream, which improves the efficiency of clustering and space usability greatly. Our framework consists of two phases. One is an online phase which approximates data streams and maintains the summary statistics incrementally. The other is an offline clustering phase which is able to perform dynamic clustering over data streams on all possible time horizons. As shown in complexity analyses and also validated by our empirical studies, our framework performed efficiently in the data stream environment while producing clustering results of very high quality.

Keywords: Clustering over evolving data streams, time series data, recent-biased approximation, data mining.

1 Introduction

A growing number of real world applications deal with multiple evolving data streams: performance measurements in network monitoring, call detail records (CDRs) in telecommunications network, performance measurements sent by distributed sensors in a sensor network, real-time quote data in the stock exchange market, and so on. Therefore, data stream mining is one of the important and challenging topics in data mining [1, 2, 3].

Clustering problem has been studied for data stream mining recently [4, 5]. In this paper, we aim to provide an approach for clustering over multiple evolving data streams on all possible time horizons. According to users' clustering request, similar data substreams are partitioned into the same clusters. A simple example is illustrated in Fig. 1. All of the four data streams from V_1 to V_4 are segmented into substreams in time horizons w_1 and w_2, and these substreams are clustered in each time horizon respectively based on similarity between each other. As a

D. Richards and B.-H. Kang (Eds.): PKAW 2008, LNAI 5465, pp. 12–26, 2009.

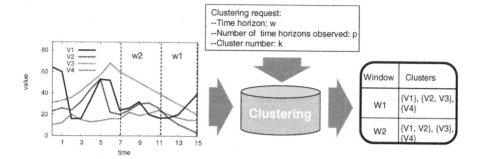

Fig. 1. Clustering of multiple data streams by our approach at $t_{now} = 15$, for a clustering request of $w = 5$, $k = 3$, and $p = 2$

result, by observing changes of clusters in consecutive user-defined time horizons, it is very efficient to identify trends and detect changes of evolving data streams.

In data streams applications, the continuous data samples are far huge to be stored in main memory generally. Therefore, the difficulty for our flexible clustering on all possible time horizons is that raw data samples of data streams cannot be revisited due to prohibitively expensive random secondary storage access. This prompts a need for an approach to maintain small-space summary statistics of each data stream on the fly.

In this paper, we develop a recent-biased approximation scheme for constructing a multi-level summary hierarchy to maintain the summary statistics of a data stream. In our approach, we provide a flexible incremental weighted approximation of data samples according to their "age" referring to the current time when a clustering request is submitted. The potential utility of such a recent-biased approximation can be found in many real world situations. For example, when analyzing real-time quote data in the stock exchange market, we consider the correlation between recent data is much more stronger, and the recent data are more significant for us to predict and make decisions than the details of ancient data. Additionally, such an approximation improves space usability greatly in the context of space constraint in data stream environment. Meanwhile, all of the approximation statistics of a data stream are maintained in a multi-level summary hierarchy. Taking advantage of the summary hierarchy, we realize the flexible clustering of arbitrary time horizons without re-approximate the original data streams.

Our approach of clustering over evolving multiple data streams is a framework consisting of an online recent-biased statistics maintenance phase and an offline clustering phase. The online recent-biased statistics maintenance phase maintains summary hierarchies of all data streams incrementally. The offline clustering phase uses these summary statistics in conjunction with other users' input in order to provide users with a quick understanding of the clusters whenever required. For example, the clustering request in Fig. 1. Since the offline phase applies clustering algorithm to the statistic rather than to the original data samples, it turns out to be very efficient in practice. This two-phases framework provides the user with the flexibility to explore the evolution of the clusters over different time horizons. This provides considerable insights to users in real applications.

The rest of this paper is organized as follows. In Sect. 2, we discuss the related work for incremental summarization and clustering over data streams. In Sect. 3, we elaborate the online recent-biased statistics maintenance phase, including how to process an incoming data incrementally and how to maintain the summary hierarchy of a data stream. In Sect. 4, we discuss how to use the summary hierarchies of data streams in the offline clustering phase in order to create clusters on arbitrary time horizons. Section 5 gives complexity analysis of our framework. Then, empirical studies are conducted in Sect. 6. This paper concludes with Sect. 7.

2   Related Work

2.1   Incremental Summarization

Several techniques have been proposed in the literature for approximation of time series, including *Singular Value Decomposition (SVD)* [6], *Discrete Fourier Transform (DFT)* [7], *Discrete Wavelet Transform, (DWT)* [8], *Piecewise Linear Approximation (PLA)* [9], and so on. Our work focuses on time-series data stream: in this situation, we cannot apply approximation techniques that require the knowledge about entire series, such as *SVD*. When considering the approximation of data samples weighted by their "age", it is not obvious that some of approximation techniques can accommodate the requirement in this new environment. *DWT* representation is intrinsically coupled with approximating sequences whose length is a power of two, and the approximation of each data sample cannot be weighted by its "age". Although *DFT* has been successfully adapted to incremental computation [10], it can neither be adapted to incremental hierarchy maintenance, since each *DFT* coefficient corresponds to a global contribution to the entire time horizon. In contrast to the above, *PLA* offers several desirable properties for the task at hand. Because each segment is independent of each other, we can reduce the fidelity of old segments simply by merging them with their neighbors, without affecting newer segments.

Bulut and Sigh proposed the usage of wavelets to represent "data streams which are biased towards the more recent values" [11]. The recent-biased concept is dictated by the hierarchical nature of the wavelet transform. In our experimental results, we compare the results of clustering on our recent-biased multi-level summary statistics and on wavelet-based approximation.

2.2   Data Stream Clustering

There are increasing number of works on dealing with clustering of data streams [4, 5, 12]. The study in [4] did not consider "evolution" of a data stream, and assumed that the clusters are to be computed over the entire data stream. Therefore, the previous results of clustering have a significant impact on the current clustering. A framework of clustering in [5] was proposed to solve the problem of clustering evolving data streams. Unlike above related works, the objective of this paper is to partition data streams or variables, rather than their data samples, into clusters. In addition, we aim to observe the evolution of the similarity between data streams within consecutive time horizons, rather than at just the current time point. On the other hand, although the objective of the study [12] is to partition data streams, it does not support clustering with flexible time horizons.

In our framework, at each time point, data samples from individual data stream arrive simultaneously. Given an integer n, an *n-stream* set is denoted as $\Gamma = \{S_1, S_2, \ldots, S_n\}$, where S_i is the i-th stream. A data stream S_i can be represented as $[S_i[1], \cdots, S_i[t], \cdots]$, where $S_i[t]$ is the data value of stream S_i arriving at time point t. $S_i[t_s, t_e]$ denotes the substream from time t_s to t_e of the data stream S_i. The objective of our approach is that given a set of data streams Γ, clustering results over multiple data streams within all possible time horizons are reported online.

3   Online Recent-Biased Statistics Maintenance Phase

3.1   Multi-level Summary Statistic Hierarchy

A Multi-level summary statistic hierarchy is proposed in the online recent-biased statistics maintenance phase in order to maintain the statistics of each data stream. In terms of space and time constraints in data streams environment, clustering algorithm is applied to statistics rather than to original data samples of data streams. In our framework, we abstract appropriate substreams within users specified time horizon from the multi-level summary statistic hierarchies of data streams. The multi-level summary statistic hierarchy plays an important role for our flexible clustering objective. Our framework realizes the abstraction of desired substreams within arbitrary observed time horizons directly from the existing summary statistic without parsing raw data samples again. In this subsection, we discuss about how to construct the multi-level summary statistic hierarchy. In the next offline clustering phase, how to abstract substreams will be discussed in detail.

Figure 2 describes the procedure of generating and updating the summary statistic hierarchy. When a new data sample arrives, it is summarized into the temporary bucket firstly according to the IncWA algorithm (which approximates data samples into K segments and will be elaborated later). Until the number of data samples in the temporary bucket reaches B_t, the coefficients of K segments as the result of approximation of B_t data samples in temporary bucket will be stored in a new basic window at level 0. The time point $t_{[0,j,i]}$ ($i <= K$) which means the end time point of i-th approximation segment in the j-th basic window at level 0 is also recorded. When B_h new basic windows are accumulated at level $(H-1)$, a new j-th basic window at level H within the sliding window

Procedure of the online recent-biased summary statistics hierarchy maintenance component (for one data steam)

1. For each incoming data example, approximate the data samples in temporary bucket incrementally according to IncWA algorithm.

2. If the number of examples in the temporary bucket is not greater than B_t, go to Step 1. Else:

    2.1 Approximation results of B_t data samples in temporary bucket is stored in a new basic window at level 0.

    2.2 Record the end time $t_{[0,j,i]}$ of segment i ($i<=K$) in the new (j-th) basic window at level 0.

3. For each level H-1 (H >1), if B_h new basic windows are accumulated:

    3.1 A new basic window at level H will be generated by incremental approximation of the approximation data in the latest B_h basic windows at level H-1 according to the IncWA algorithm.

    3.2 Record the end time $t_{[H,j,i]}$ of segment i ($i<=K$) in the new (j-th) basic window at level H.

Fig. 2. Procedure of online recent-biased maintenance phase

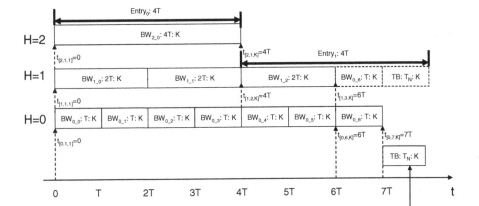

Fig. 3. The illustration of summary hierarchy, where $BW_{1\_0} : T : K$ represents the 0-th basic window at level 1 of time horizon $B_t = T$, K is the number of segments used for approximation of a basic window, $t_{[1,1,K]}$ records the end time point of the K segments of the first basic window at level 1 and $B_h = 2$

$[t_{[H-1,B_h(j-1),0]}, t_{[H-1,(B_h j),K]}]$ begins to be generated. The IncWA algorithm begins the approximation process of generation, and approximation data generated from the $B_h K$ approximation segments in the above sliding window at level $(H-1)$ are used. A sample of summary statistic hierarchy is illustrated in Fig. 3. The parameters $B_t = T$, $B_h = 2$, and K is the number of approximation segments used for approximation of each basic window.

A summary hierarchy is established for each data stream. Therefore, totally n hierarchies are maintained for n data streams. Due to the space limitation in the streaming environment, only the latest α basic windows are maintained at each level. Note α should be set to a number not smaller than B_h in order to keep enough basic windows at lower level for the generation of a basic window at a higher level.

3.2   Incremental Recent-Biased Approximation

In the context of data streams, where data samples are continuously generated, potentially forever, we take the advantage of PLA [9] for linear and one single scan of streams. In PLA, a raw data stream is approximated by linear segments incrementally. The linear estimation function $\hat{S}[t] = b + at$ for an approximation segment s corresponding to substream $S[t_s, t_e]$ is taken to be the best approximation line conforming to the principle of least squares. The error of the segment s is computed according to the formula $E(s) = \sum_{t_j \in S}(S[t_j] - \hat{S}[t_j])^2$, where t_j ranges over $[t_s, t_e]$.

As mentioned earlier, in some applications the usefulness of a data sample can diminish with its "age". Therefore, it is not appropriate to approximate old data samples as accurate as recent ones. We provide a flexible and incremental recent-biased approximation of data samples in a basic window. We define the weight function of a previous data sample $S[t_j]$ as $A(x)$, where $x = t_{now} - t_j$ is its "age" referring to the arrival of an incoming data sample, t_{now} is the current time point at which a new data sample arrived, and t_j is the time point at which $S[t_j]$ arrived.

We specifies the number of segments to approximate data samples within a basic window equals to K. Statistics of the basic window $[t_s, t_e]$ consists of coefficients of the K segments $(s_1, s_2, \ldots, s_K)$. We do incremental statistics collection at each time point within sliding window $[t_s, t_e]$ in order to minimize the relative approximation error $RE(S[t_s, t_e]) = \sum_{j=1}^{K}(E(s_j)/A(t_e - t_j))$, where t_j is the time point at which the most recent data sample of segment s_j arrived.

Our algorithm for incremental recent-biased approximation of a basic window is abbreviated as IncWA algorithm. In order to minimize the relative approximation error, the IncWA algorithm merges the two consecutive segments whose mergence will result in the least approximation error, among all possible mergences. A queue structure Q is used to determine the order of mergence of two consecutive segments. Assume that the segments of the first pair in Q are s_m and s_{m+1}, we merge those into one segment $s_{m,m+1}$. Then we compute the approximation errors that would result from merging the new segment with its two neighbors, s_{m-1} and s_{m+1}. We update Q, in order to reflect a new set for possible merge. A skeleton description of the algorithm is depicted in Fig. 4. The IncWA algorithm also makes use of a time points queue TQ. This structure keeps track of the way that the dependencies among the segments used for the approximation change as a result of the amnesic function. These dependencies are managed in the procedure ManageEvents.

Each time point in TQ specifies crosspoint at which the relative merge ordering of two pairs of segments changes. Suppose that t_{s1} and t_{s2} are time points at

```
Let TQ = NULL be a time points queue;
Algorithm IncWA
INPUT:  S[tᵢ]: a new data sample
        Q: queue
        TQ: event time points queue
OUPUT: K segments to approximate data stream
  1.    when a new data sample S[tᵢ] arrives, assign a new segment, s' to the newly arrived example S[tᵢ];
  2.    ManageEvents (TQ, tᵢ);
  3.    pick the minimum element from Q, corresponding segments, sₘ and sₘ₊₁;
  4.    compare the mergence error of sₘ and sₘ₊₁ with the mergence error of s' and its neighboring segment;
  5.    merge the corresponding segments with less mergence error;
  6.    update Q with the error of merging s' with its neighboring segment;
  7.    update Q with the errors of merging sₘ,ₘ₊₁ with its two neighboring segments;
  8.    ManageEvents (TQ, tᵢ);
  9.    return;

Proc ManageEvents (queue TQ, time tᵢ)
  1.    if (next crosspoint tₑ in TQ satisfies tᵢ < tₑ <= tᵢ+1)
  2.       then remove tₑ, related to segments sₑ,₁ and sₑ,₂;
  3.          change in H the positions of sₑ,₁ and sₑ,₂;
  4.          compute crosspoint between sₑ,₁ and sₑ,₂ with all other segments in Q;
  5.          insert in TQ time points for any new crosspoint identified;
  6.    insert in TQ any new crosspoint identified concerning every pairs of segments in Q;
  7.    return;
```

Fig. 4. IncWA algorithm

which the most recent data samples of segments s_1 and s_2 arrived respectively: their crosspoint, t_c is given by the following equation, where $A(x) = x$.

$$\frac{E(S[t_e - t_{s1}])}{t_c - (t_e - t_{s1})} = \frac{E(S[t_e - t_{s2}])}{t_c - (t_e - t_{s2})}$$

4   Offline Clustering Phase

In the offline clustering phase, we aim to provide dynamic clustering over multiple evolving data streams on all possible time horizons. Note that the desired time horizon would be different from the size of basic windows maintained in the summary statistic hierarchy. In this situation, we have to select the basic windows from appropriate levels of the hierarchy to generate the substream with the length of user-specified time horizon. Here we define an *entry* as the substream which covers the user-specified time horizon. When user's clustering request of time horizon w is submitted, at most p entries from each data stream will be observed. We have the following definitions for appropriate level selection.

Definition 1. *The highest level to approximate an entry with time horizon w is defined as*

$$H_{max} = \lfloor \log_{B_h} (\frac{w}{B_t}) \rfloor \tag{1}$$

Definition 2. *The lowest level to approximate an entry with time horizon w is defined as*

$$H_{min} = \min_H \{w \le (t_N - t_{[H,i,K]}) + \alpha_H h_H\} \tag{2}$$

Procedure of offline clustering phase ()

1. Calculate H_{min} and H_{max}. For each data stream do Steps 2 and 3.

2. If the end time point $t_{[Hmin, i, K]}$ of level H_{min} is not equal to the current time, aggregate the basic windows of lower levels (from $(H_{min} -1)$ to 0) and segments of the temporary bucket to generate a temporary basic window to characterize the interval between $t_{[Hmin, i, K]}$ and the current time. Then, aggregate this temporal basic window to the latest basic window in level H_{min}.

3. Encapsulate the basic windows between level H_{min} and H_{max} to generate at most p entries. Set $H = H_{min}$ initially. For the entries from w_1 to w_p, if the range of a desired time horizon is covered by the interval of the basic windows at level H, encapsulate an approximate number of basic windows into that entry, Else, increase H by one to look for the basic windows with enough coverages. This step stops when p entries have been retrieved or when H exceeds the maximum level H_{max} with p_r entries obtained, where $p_r <= p$.

4. For w_1 to w_p, run clustering algorithm to cluster entries of n data streams.

Fig. 5. The outline of the offline clustering phase

where α_H is the exact number of basic windows at level H, $t_{[H,i,K]}$ is the end time point of K-th segment of the most recentest basic window (i-th basic window) at level H before the clustering request which is submitted at t_N, and $h_H = B_t B_h{}^H$ is the size of a basic windows at level H.

From the above definitions, basic windows at the levels between H_{max} and H_{min} are used to approximate entries, and at most p entries of the time horizon w can be retrieved from each data stream. The outline of the offline clustering phase is shown in Fig. 5. In the first step the values of H_{min} and H_{max} are calculated according to the time horizon w specified by the clustering request. Then, entries which cover the ranges of the desired time horizon, are retrieved from each data stream by Step 2 and Step 3. In Step 3, we encapsulate the basic windows at one level of the hierarchy which cover the range of the desired time horizon into an entry. Finally, for each time horizon the clustering algorithm is executed on the retrieved entries.

Let us consider Step 2 in the offline clustering phase. Assume the level $H_{min} = 2$, as shown in Fig. 3. Since the end time point $t_{[2,1,K]}$ is not equal to the current time, we aggregate the basic windows from lower levels (from level 1 to 0) and segments from the temporary bucket to generate a temporary basic window which characterize the interval between $t_{[2,1,K]}$ and the current time. Then, it is aggregated into the latest basic window of level 2.

For each data stream, the r-th ($r \leq p$) entry with the length of time horizon w at level $H'(H_{min} \leq H' \leq H_{max})$ can be recorded by $(2K\lfloor \frac{w}{B_t \times (B_h{}^{H'})} \rfloor)$ statistics. Therefore, the *distance* between two entris $S_i(w_r)$ and $S_j(w_r)$ of window w_r can be calculated by Euclidean distance of these statistics.

5 Complexity Analysis

The complexities of the online and offline phases in our framework are shown in the following theorems, where n is the number of streams and m is the number of data samples in each stream.

Theorem 1. *The space complexity of the IncWA algorithm of a basic window is $O(K)$, and the time complexity to process each new data sample is O (logK).*

Proof. The algorithm requires $O(K)$ space to store K segments and the segments queue Q, and $O(TQ)$ space for the time points queue. In practice, the size of TQ remains small. At each iteration, time to find a pair of segments for merging, and two pairs of segments which reach a crosspoint, is $O(1)$. We need $O(\log K)$ time to update Q after above changes. We also need to update TQ, which takes $O(\log |TQ|)$ time, and it can be omitted since the size of TQ remains small. Therefore, the overall time complexity for each iteration is $O(\log K)$.   □

Theorem 2. *The time complexity of the online recent-biased summary statistic maintenance phase is $O\left(n(m + \frac{Km}{B_t})\log K\right)$.*

Proof. For each data stream, in level 0, $\frac{m}{B_t}$ basic windows can be generated in time $O(m)$ after m data samples arrive. In each basic window, B_t times iteration will be processed to generate K segments: according to Theorem 1, we can get the total time complexity at level 0 is $O(m \log K)$. The aggregation of B_h basic windows at lower level and generation of a new basic window at higher level requires $2KB_h$ iteration of IncWA algorithm, That is, the complexity is $O(KB_h \log K)$. Accordingly, totally the number basic windows of all levels are

$$\frac{m}{B_t} \times \left(\frac{1}{B_h} + \frac{1}{B_h{}^2} + \cdots + \frac{1}{\log_{B_h}(\frac{m}{B_t})}\right) \le \frac{m}{B_t} \times \left(\frac{\frac{1}{B_h}}{1 - \frac{1}{B_h}}\right) = O\left(\frac{m}{B_t(B_h - 1)}\right). \text{ Time for}$$

building the $O\left(\frac{m}{B_t(B_h-1)}\right)$ basic windows will be $O\left(\frac{K \log Km}{B_t}\right)$. Consequently, the overall time complexity of the online recent-biased statistics maintenance phase for n streams is $O\left(n(m + \frac{Km}{B_t})\log K\right)$.   □

Theorem 3. *The space complexity of the online recent-biased summary statistic maintenance phase is $O\left(n\alpha K \log_{B_h} \frac{m}{B_t}\right)$.*

Proof. For each data stream, in level 0, $\frac{m}{B_t}$ basic windows have been created after m points arrived. B_t basic windows at lower level are aggregated to a basic window at higher level. Accordingly, the height of the summary hierarchy is $\log_{B_h}(\frac{m}{B_t})$ and each level maintains at most α basic windows. Note that each basic window contains a constant number of parameters. Consequently, totally $n\left(\alpha \log_{B_h} \frac{m}{B_t}\right)$ basic windows are maintained for n streams. According to Theorem 1, the space complexity of one basic window is $O(K)$. Therefore, the space complexity of the online recent-biased summary statistic phase is $O\left(n\alpha K \log_{B_h} \frac{m}{B_t}\right)$.   □

Theorem 4. *The space complexity of the offline clustering phase is $O(kn\alpha)$ for one entry.*

Proof. The complexity of k-means clustering algorithm is $O(knd)$, where d is the number of dimensions. Since the time horizon in a clustering request is represented by at most α basic windows with constant number of parameters. Therefore, the time complexity of the offline clustering phase is $O(kn\alpha)$ for temporal granularity.

□

6 Experimental Results

Our experiments focus on two aspects: 1) Is our two-phased framework suitable for clustering of data streams? Does our framework partition data streams well? We do sensitivity and scalability analysis to discuss partition ability, execution time and scalability of our framework. 2) Is our framework able to result appropriate clusters in terms of similarity of data streams? For this purpose, we look into the cluster results of our proposed approach, and compare them with Euclidean distance based clustering over raw data streams without considering about the "age" of data samples.

6.1 Experiments Setup

We use S&P stock data for experiments [13]. We collected a one-year historical data set over 476 stocks. We use the high value of each stock with the same length of 1252.

Our task is to mine out clusters of stocks within user-specified time horizons. As one of the evaluation of our framework's partition ability, we use the average intra-cluster distance as the evaluation function for clustering results:

$$\text{IntDis}(Cl) = \frac{\sum_{i=1}^{p} \text{IntDis}(Cl(w_i))}{p}, \qquad (3)$$

where

$$
\begin{aligned}
&\text{IntDis}(Cl(w_i)) \\
&= \frac{\sum_{C_j(w_i)} \sum_{S_q(w_i) \in C_j(w_i)} dist\left(S_q(w_i) - C_j(w_i).center\right)}{n \times m},
\end{aligned}
$$

$C_j(w_i).center$ is center of cluster C_j in entry w_i.

In the following experiments, the average intra-cluster distances of our framework are measured as the ratio of that obtained by running the k-means clustering algorithm on the raw data streams. In order to reduce the impact of initialization of k-means clustering algorithm, we take the average of 20 times experiment results. Without loss of generality, we assume that the bucket size $B_t = B_h = B$.

6.2 Sensitivity Analysis

In this section, two parameters of the summary hierarchy, which are the number of basic windows maintained at each level α and bucket size B are investigated on the stock data with a fixed window size $w = 50$. Because the clustering time is mostly dominated by the execution time of the k-means clustering algorithm, as shown in Fig. 6(a) and 6(b), α does not have much influence on the average intra-cluster distances and execution time of clustering. As shown in Fig. 6(c) and 6(d), it is clearly that when the bucket size is too small, the execution time of our framework will be larger. We can find that clustering on the summary hierarchies of our framework is able to achieve almost the same quality as that on raw data streams efficiently.

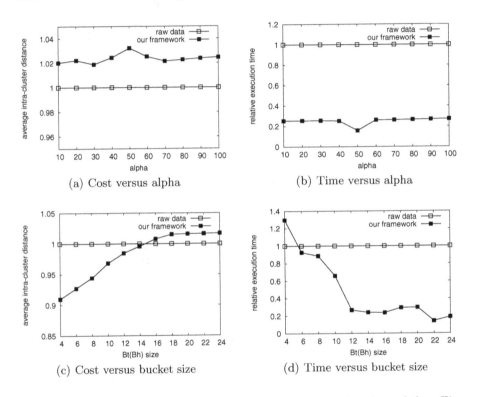

Fig. 6. The average intra-cluster distance and relative execution time of the offline clustering phase of our framework

The performance of our framework on real clustering requests is next evaluated with a fixed bucket size $B = 12$. As shown in Fig. 7(a) and 7(c), while varying the size of time horizon w and the number of entries observed p, our framework is able to obtain the same good clustering quality as the clustering on the raw data streams, with significantly shorter execution time, as shown in Fig. 7(b) and 7(d).

6.3   Scalability Analysis

To evaluate the scalability of the online recent-biased summary statistics maintenance phase, scale-up experiments on both the number of data points m and

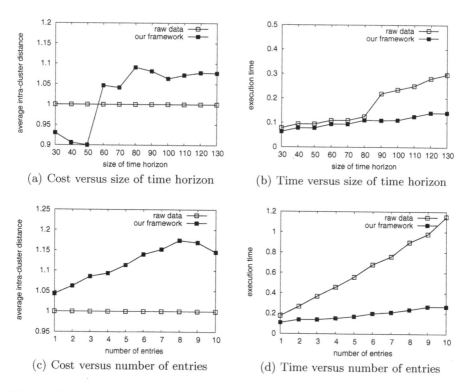

(a) Cost versus size of time horizon   (b) Time versus size of time horizon

(c) Cost versus number of entries   (d) Time versus number of entries

Fig. 7. The average intra-cluster distance and execution time of the offline clustering phase of our framework

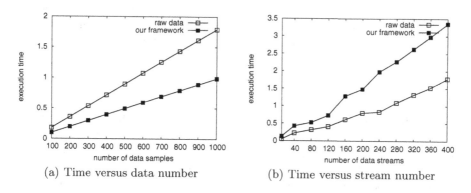

(a) Time versus data number   (b) Time versus stream number

Fig. 8. The scalability analyses of the online recent-biased summary statistics maintenance phase of our framework

the number of data streams n are conducted. As shown in Fig. 8(a), as the number of data points in each stream increases from 100 to 1000, the execution time grows linearly. This property also holds when the number of data streams varies

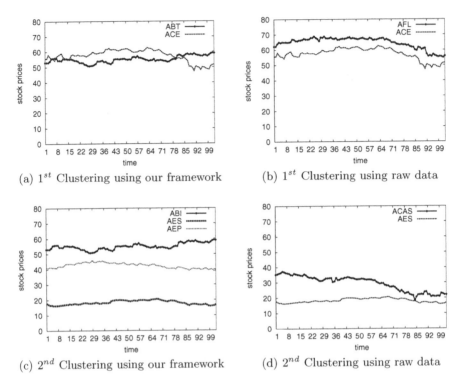

(a) 1st Clustering using our framework   (b) 1st Clustering using raw data

(c) 2nd Clustering using our framework   (d) 2nd Clustering using raw data

Fig. 9. Comparison of results of clustering using our framework and real data streams

from 40 to 400, as shown in Fig. 8(b). These results conform to our analyses in Sect. 5 which states that the time complexity of the online summary statistics hierarchy maintenance phase of is linear to both the number of streams and the number of data samples in each stream.

6.4   Cluster Analysis

Figure 9 shows two sets of clusters we found out using our framework and real data streams. In the 1st result set as shown in Fig. 9(a) and 9(b), although the two stock data in Fig. 9(a) have opposite trends rather than similar trend, average trends of these two data stream are the same. Our framework mined this kind of cluster. In many applications, this kind of opposition information is also interesting. On the other hand, Euclidean-based clustering on raw data streams cannot mine the opposition information in Fig. 9(b). In the 2nd result set, our framework partition three similar data streams into a cluster as shown in Fig. 9(c), while the "AES " stock data was partitioned into different cluster in the clustering on raw data streams as shown in Fig. 9(d). From this result, we can see that our method for summary statistics collection can do similarity search invariant to shifting, while Euclidean distance cannot reflect the real similarity between data streams in the case of shifting.

7   Conclusions

According to the effective integration of the online summary statistics maintenance phase and the offline clustering phase, our framework realizes: (1) incremental and recent-biased summarization of statistics, is helpful to discover interesting recent-biased correlations between data streams, (2) online maintenance of statistics in multi-level hierarchies, is helpful for flexible clustering on all possible time horizons efficiently, (3) Clustering over consecutive time horizons is helpful to observe the evolution of clusters and identify the trend of data streams. As shown in the complexity analyses and also validated by our empirical studies, our framework performed efficiently in the data stream environment while producing clustering results of very high quality. In our future work, we are interested in discovering lagged-correlated patterns by clustering arbitrary aligned substreams of multiple evolving data streams.

References

1. Aggarwal, C.C., Han, J., Wang, J., Yu, P.S.: On Demand Classification of Data Streams. In: 10th ACM SIGKDD International Conference on Knowledge Discovery and Data mining, pp. 503–508. ACM Press, Seattle (2004)
2. Metwally, A., Agrawal, D., Abbadi, A.E.: Efficient Computation of Frequent and Top-k Elements in Data Streams. In: Eiter, T., Libkin, L. (eds.) ICDT 2005. LNCS, vol. 3363, pp. 398–412. Springer, Heidelberg (2004)
3. Gaber, M.M., Zaslavsky, A., krishnaswamy, S.: Mining Data Streams: A Review. ACM SIGMOD Record 34(2), 18–26 (2005)
4. O'Callaghan, L., Mishra, N., Meyerson, A., Guha, S., Motwani, R.: Streaming-Data Algorithms for High-Quality Clustering. In: 18th International Conference of Data Engineering, pp. 685–694. IEEE Press, San Jose (2002)
5. Aggarwal, C.C., Han, J., Wang, J., Yu, P.S.: A Framework for Clustering Evolving Data Streams. In: 29th International Conference on Very Large Data Bases, pp. 81–92. Morgan Kaufmann, Berlin (2003)
6. Chakrabarti, K., Keogh, E.J., Mehrotra, S., Pazzani, M.J.: Locally Adaptive Dimensionality Reduction for Indexing Large Time Series Databases. ACM Trans. Database Systems 27(2), 188–228 (2002)
7. Rafiei, D.: On Similarity-Based Queries for Time Series Data. In: 15th IEEE International Conference on Data Engineering. IEEE Press, Sydney (1999)
8. Popivanov, I., Miller, R.J.: Similarity Search Over Time Series Data Using Wavelets. In: 18th IEEE International Conference on Data Engineering, pp. 802–813. IEEE Press, San Jose (2002)
9. Keogh, E.J., Pazzani, M.J.: An Enhanced Representation of Time Series Which Allows Fast and Accurate Classification, Clustering and Relevance Feedback. In: 2nd International Conference on Knowledge Discovery and Data Mining, pp. 239–243. ACM Press, Montréal (1998)

10. Zhu, Y.Y., Shasha, D.: StatStream: Statistical Monitoring of Thousands of Data Streams in Real Time. In: 28th International Conference on Very Large Data Bases, pp. 358–369. Morgan Kaufmann, Hong Kong (2002)
11. Bulut, A., Singh, A.K.: SWAT: Hierarchical stream summarization in large networks. In: 19th IEEE International Conference on Data Engineering, pp. 303–314. IEEE Press, Bangalore (2003)
12. Yang, J.: Dynamic Clustering of Evolving Streams with a Single Pass. In: 19th International Conference of Data Mining, pp. 695–697. IEEE Press, Bangalore (2003)
13. http://kuo.swcp.com/stocks/

Automatic Database Creation and Object's Model Learning

Nguyen Dang Binh and Thuy Thi Nguyen

Institute for Computer Graphics and Vision
Graz University of Technology, Austria
{binh,thuy}@icg.tugraz.at

Abstract. This paper proposes a new framework to automatically generate visual object database meanwhile efficiently learn the object's model. The system is of important need for the problems of object detection and recognition. Our main idea is to acquire the huge amount of video data actively, and seeks out opportunities to autonomously exploit information from object samples. We employ autonomous learning approach based on online boosting technique, which allows to combine an object detector trained on a single initialized input image with tracking to extract object samples for learning. The autonomous learning process with interactive learning strategy allows to adaptively improve the learning object model while generating informative samples. Our method allows to generate thousands of object samples within hours from large video databases or from live camera, thus saving time and labor's efforts. We will show that the proposed method can extracts well-localized, diverse appearances of object examples from video sequence through only one initialized input sample, and builds robust object model. In addition to requiring very little human intervention, a significant benefit of this method is that it does not require pre-training. In the experiments, the approach is evaluated in detail for creating data sets and learning for the problems of human hand gesture recognition and face detection. In addition, to show the generality, results for different objects are also presented.

Keywords: Database creation, object's model learning, online boosting, autonomous learning, object detection, pattern recognition.

1 Introduction

In the last decade, computer vision has been being a fast growing research field. A lots of researches have been focused on acquiring knowledge about visual object of interest by training of detectors but only a little attention has been paid to efficiently labeling and acquiring suitable training data. Training a reliable object model requires large dataset, where positive and negative samples are usually obtained by hand labeling from large number of images. This costs a lots of time and labor efforts. In addition, the construction of appearance-based object detection systems is challenging because a large number of training examples must be collected and manually labeled in order to capture variations in object appearances. The number of variables that may

D. Richards and B.-H. Kang (Eds.): PKAW 2008, LNAI 5465, pp. 27–39, 2009.
© Springer-Verlag Berlin Heidelberg 2009

be relevant in the database distribution is immense such as viewing angle, scale variation, lighting condition, background clutter, etc. With so many variables, there is no assurance that the training database has taken all the relevant variables into account or that their distribution will be the same as will be found in the application context.

The popular method for training a visual object classifier is to use a supervised learning algorithm (e.g. Adaboost [3, 4], neural network [5], or support vector machine [6]) with large hand-labeled set of object and non-object images. Most of these approaches have some drawbacks: First, at the preparation phase, manual labeling of training data is mandatory. For reliably training an object detector a large training data, in the order of several thousand image patches [7], is required. Labeling such a large training set by hand is costly. For the scope of many projects, creating a training set of over a thousand images is unrealistic. It is important to keep the human supervisory effort to a practical level. Second, for offline learning strategy, learning process is performed offline before the classifier is used for detection/recognition task. So, after the training the classifier remains fixed, and any further training is not possible. However incremental, online learning is desirable because then the classifier needs to know only what is actually necessary for the specific task. The classifier is time-adaptive and online learning can continue as long as the task is performed.

Recent years, online boosting learning [1, 17] has become widely used in computer vision community. Although online boosting allows efficiently training and improving the detector with a small training data set, learning an object detector is performed by an interactive process of just clicking to select positive and negative object samples on the current image. The evaluation the current classifier is done at a time. Thus the classifier can not update temporal information that represent contiguous variances of object samples, i.e. an object moving under camera or video sequences (e. g hand gesture or body gestures).

Obtaining a set of rich and informative training samples is challenging problem, especially for positive examples. To reduce the labeling effort, there have been a number of approaches which used two phases in a co-training fashion [7, 8]. Usually, at the first phase, a classifier is trained on small set of training data. Then at the second phase, the classifier is employed to detect object patches which will be used as positive samples. [8] even completely avoid hand labeling by using motion detection to obtain the initial training set. New examples are acquired by applying a detector obtained by online learning. Negative examples (i.e. examples of images not containing the object) are usually obtaining by a bootstrap approach [9]. A drawback of these approaches is one has to train an initial classifier. Moreover, the classifier tends to be biased by the examples used to train the initial model. Thus, many potential new examples, which represent different views of the same object, may not fit to the learned model.

By combining a tracking process with learning in a framework, the variation of object appearances can be tracked through image sequences, which provide samples for learning. Using a classifier and a tracker together, we take the advantage of the temporal continuity of video sequences to validate both tracking and classification, one for the other, while generating additional training examples. Our approach does not rely on pre-trained classifiers to bootstrap the learning process. We start with only one initial input image sample.

The main contribution of this paper is an autonomous learning of object's model that overcomes the above limitations. The basic idea is to design an automatic labeler, which can be seen as an oracle, to generate databases, i.e. positive and negative training samples, meanwhile incrementally learn the object's model. We develop an autonomous online learning algorithm based on online boosting [1], which allows to update the existing classifier based on one positive sample delivered by the oracle at each iteration to learn the object model and generate the object's database.

2 Related Work

There are numerous methods for moving object detection use motion as their primary source of information. Levin et al. [7] applied motion information to reduce the hand-labeling effort. They initially trained two classifiers, one on background difference images and the other one on intensity images with a small number of hand-labeled examples. Then the two classifiers use unlabeled data to iteratively improve each other. Our approach differs from [7] in that we use an autonomous classifier as tracker instead of the second classifier to identify informative training examples. Likewise, Javed et al. [17] used co-training to improve the performance of an initial classifier by selecting new training examples using PCA. Both systems needed a non negligible amount of supervision for labeling during initial training. Sivic et al. [13] applied tracking to obtain training samples. They used a face detector [14] that is trained by boosting orientation-based features. A conservative detection threshold is used to obtain low false positive rate. The consequence is many faces are not detected and the false negative rate is increased. [12] initialized an affine covariant region tracker to compute face representation from the tracked patches. First, they searched to localize facial features such as eyes, tip of the nose, and the center on the mouth. Then the object representation is built from five overlapping SIFT descriptor at the detected features. The drawback is to learn the model for the feature position and appearance, thousands hand-labeled face images is needed. The strength of the learned classifier-based detection approach is that it selects the object model using a learning algorithm, based explicitly on the model's ability to discriminate between object and non-object training examples. Hewitt and Belingie [15] proposed a method to learn a face representation, where a tracker serves for verification. The tracker locates the face correctly whereas the initial classifier may fail. Wu et al. presented an approach to online (re)-training of a detector based on the output of a coarse detector using boosting. As boosting focuses on difficult examples during training, it may be unstable if some examples are wrongly labeled. The method [6] and [7] also need supervision for initial stages, and it can only learn objects having the appearance similar to the samples used in initial training. Most of these mentioned approaches have been applied in one context only (e.g. pedestrian or car detection).

Several limitations should be addressed are: First, a pre-trained classifier is needed to initialize the learning process. Besides, a simple tracker may make some errors and select wrong samples, which must be verified manually before feeding to the learning process. Hence, non negligible human supervision effort is necessary. Second, tracker provides the labels during tracking, which would allow online learning, still the models are trained off-line. In addition, there is no verification process on samples learned so far. Finally, there is no attempt to collect generated samples as the automatic acquisition of training data to build the objects databases.

In this paper we introduce an autonomous learning framework that based on online boosting learning. It requires no initial or pre-training. The tracker is initialized once, e. g with a good positive sample on the first single input image. Afterward, no user inter-action is needed. In particular we employ online boosting for both learning and track-ing, which allows to learn online an object detector and generate object samples. The idea is to use tracking information for selecting the most valuable positive and negative samples. An existing classifier is directly updated and evaluated on current image. The thus obtained detection results are the true positives and false positives as negatives. These samples are used to update the classifier and stored to the object's database. This proposes a simplification of the sample's generation process, in which a computer can train itself to detect and distinguish individual objects. It is a mean for reducing human effort needed to prepare the training set by training the object model. The process can perform real-time for processing images from a live camera or video sequence.

The paper is organized as follows. Section 1 is given to introduction. Section 2 pre-sents the related work. Section 3 describes our framework for learning object model and generation of training data. Experiments and results are shown in section 4. Sec-tion 5 is for conclusion and future work.

3   Learning Framework

3.1. On-Line Learning

We employ the on-line boosting learning for feature selection as proposed in [1]. In the following, we briefly summarize the method. The main idea of boosting learning for feature selection is that each feature f_j corresponds to a single weak classifier h_j and that boosting selects an informative subset of N features, where a weak classifier has to perform only slightly better than random guessing (i.e., the error rate of a classifier for a binary decision task must be less than 50%). In fact, various different feature types may be applied but similar to the seminal work of Viola and Jones [3] in this work we use Haar-like features, which can be calculated efficiently using integral data-structures.

In the off-line case boosting for feature selection can be summarized as follows: given a training set of positive and negative samples $\chi = \left\{ \langle x_1, y_1 \rangle, ..., \langle x_L, y_L \rangle \big| x_i \in R^m, y_i \in \{-1, +1\} \right\}$ where $x_i \in R^m$ is a sample and $y_i \in \{-1, +1\}$ is the corresponding label, a set of possible features $F = \{f_1, ..., f_M\}$, a learning algorithm $\Im$, and a weight distribution D, that is initialized uniformly by $D(i) = \dfrac{1}{L}$. In each iteration n, $n = 1, ..., N$, all features $f_j, j = 1, ..., M$ are evaluated on all samples $(x_i, y_i), i = 1, ..., L$ and hypotheses are generated by applying the learning algorithm $\Im$ with respect to the weight distribution D over the training samples. The best hypothesis is selected and forms the weak classifier h_n and the weight distribution D is updated according to the error of the selected weak classifier. The process is repeated until N features are selected (i.e., N weak classifiers are trained). Finally, a strong classifier H is computed as a weighted linear combination of all weak classifiers h_n.

Contrary, during on-line learning each training sample is provided only once to the learner. Thus, all steps described above have to be on-line and the weak classifiers have to be updated whenever a new training sample is available. On-line updating the weak classifiers is not a problem since various on-line learning methods exist, that may be used for generating hypotheses. The same applies for the voting weights α_n that can easily be computed if the errors of the weak classifiers are known. The crucial step is the computation of the weight distribution since the difficulty of a sample is not known a priori. Thus, the basic idea is to estimate the importance λ of a sample by propagating it through the set of weak classifiers [18]. In fact, λ is increased proportional to the error e of the weak classifier if the sample is misclassified and decreased otherwise.

Thus, the work-flow for on-line boosting for feature selections selection can be described as follows: a fixed number of N selectors $s_1,...,s_N$ are initialized with random features. A selector s_n can be considered a set of M weak classifiers $\{h_1,...,h_M\}$, that are related to a subset of features $F_n = \{f_1,...,f_k\} \in F$, where F is the full feature pool. The selectors are updated whenever a new training sample $\langle x,y \rangle$ is available and the selector $s_n(x)$ selects the best weak hypothesis according to the estimated training error from the importance weights of the correctly and incorrectly classified samples seen so far. Finally, the weight α_n of the n-th selector s_n is updated, the importance λ_n is passed to the next selector s_{n+1}, and a strong classifier is computed by a linear combination of N selectors:

$$H_{on}(x) = sign\left(\sum_{n=1}^{N} \alpha_n \cdot s_n(x)\right) \qquad (1)$$

Thus, contrary to the off-line version, an on-line classifier is available at any time of the training process.

3.2 Autonomous Online Learning

We will present our autonomous online learning algorithm to learn incrementally an object's model and efficiently generate training data. The learning process begins with only one initialized example using online boosting that has been discussed in detail in Section 3.1.

First, to initialize the classifier, a selected image region is assumed to be a positive sample. We have one-click to select target object as positive sample $\langle x,+1 \rangle_{t=0}$ and

$Pos_{t=0} = Pos_{t=0} \cup \{\langle x,+1 \rangle\}$. The current target region is used for a positive update of the classifier $C_{t=0}$. Given this positive sample, an initial classifier $C_{t=0}$ is trained. The classifier is evaluated, and once the target object has been detected (the best of the detection) at time t, it is considered to be a positive image sample $\langle x,+1 \rangle_{t=1}$ for updating of the classifier. At the same time, false positives are determined and used as negative samples $\{\langle x_1,-1 \rangle,...,\langle x_n,-1 \rangle\}_{t=1}$ for update. These negative samples are obtained by taking regions of the same size as target window from the false positives in the surrounding background: $Neg_t = Neg_{t-1} \cup \{\langle x,-1 \rangle\}$.

Using these samples to update, several iterations of the online boosting algorithm are carried out. Thus the classifier adapts to the specific target object and at the same time it is discriminative against its surrounding background. At time t, the classifier C_{t-1} is applied on the current image I_t. Thus the obtained detection result is verified by the tracking result T_t that robustly represents the object-of-interest. Based on this verification, the valuable samples (see Figure 1), i.e., the reported false positives (blue bounding boxes), are identified. In addition, such selected samples are labeled. These samples are fed back into the discriminative classifier as positive and negative examples, respectively, and we get a better classifier C_t. Obviously, the number of negatives is theoretically infinite if a non-integer positive grid is used. The current C_t classifier is evaluated at the surrounding region of interest and so obtains for each sub-patch a confidence value which implies how well the underlying image patch fits the current model. Afterwards we choose the best of the detection as maximum obtained confidence and shift the target window to the new maxima location, and $Pos_t = Pos_{t-1} \cup \{\langle x,+1 \rangle\}$. Next, the classifier has to be updated in order to adjust to possible changes in appearance of the target object and to become discriminative to a different background. The current target region is used for a positive update of the classifier while surrounding false positive regions are taken as negative samples and $Neg_t = Neg_{t-1} \cup \{\langle x,-1 \rangle\}$. To cover as many negative as possible we maintain the same set of positives but bootstrap a new set of negatives that pass all previous strong classifier (i.e. false positive). This update policy has proved to allow stable learning and tracking in natural scenes. As new frames arrive, the whole procedure is repeated and the classifier is therefore able to adapt to possible appearance changes and in addition becomes robust against background clutter.

    The idea is to employ online boosting technique to adaptively learn an object representation/discriminative classifier from only one initialized example. To actually learn the object representation we develop autonomous online learning algorithm based on online boosting learning algorithm [1] but any other online learning method my be applied.

Algorithm 1. Online Autonomous Learning and Data Generation

Input: - An empty discriminative classifier $C_{t=0}$;

      - Video sequence or image set

Output: Classifier C_t; Positive set Pos_t and Negative set Neg_t and Ground truth;

1: Initialize parameters for the classifier $C_{t=0}$ and train with 1-click on initial object sample;

2: Initialize positive and negative sets: $Pos_{t=0} = \{\}$ and $Neg_{t=0} = \{\}$

3: **while** Non-Stop-Criteria **do**

4: Evaluate C_{t-1} on current image frame I_t obtain J detection x_j and display results;

5: Predict and determine true positive: T_{t-1};

6: **For** j=1,..., J **do**

    If $T_{t-1} \approx x_j$ then

        begin

            //Use true positives samples to update the classifier;

            ●Update(C_{t-1}, x_j,+1) follows algorithm 2;

            // Automatic true positive labeling: adding true positive to Pos_t set ;

        ● $Pos_t = Pos_{t-1} \cup \{\langle x,+1\rangle\}$;

        end

    Else $T_{t-1} \neq x_j$ then   //Determine false positives on current image I_t ;

        begin

            //Use false positives as negative samples to update the classifier;

            ●Update(C_{t-1}, x_j,-1) follows algorithm 2;

            // Automatic negative labeling: adding negative samples to Neg set ;

        ● $Neg_t = Neg_{t-1} \cup \{\langle x,-1\rangle\}$;

        end

7: **End for**

8. **End while**

3.3 Image Representation and Features

In our work, we use efficient integral image representation for fast calculation of objects features. The features include Haar wavelet [3], local orientation histogram [19] and a simplified version of local binary patterns [20] as a representation, which can be fast computed on integral images. The computation of these feature types can be done very efficiently. For online learning a weak classifier h_j for feature j we

---

Algorithm 2. Online Learning for Feature Selection

Input: - Training example $\langle x, y \rangle$, $y \in \{-1,+1\}$;

   - Strong classifier C_{t-1};

   - Initialized weight $\lambda_{n,m}^{corr} = 1$; $\lambda_{n,m}^{wrong} = 1$;

Output: Strong classifier C_t

1: Initialized the importance weight $\lambda = 1$

2: **For** n=1,...,N **do** // for all selectors

   for m=1,...., M **do** //update the selector s_n^{sel}

   - $h_{n,m}^{weak} = update(h_{n,m}^{weak}, \langle x, y \rangle, \lambda)$; // update each weak classifier

   - if $h_{n,m}^{weak}(x) = y$ then $\lambda_{n,m}^{corr} = \lambda_{n,m}^{corr} + \lambda$;

      else $\lambda_{n,m}^{wrong} = \lambda_{n,m}^{wrong} + \lambda$;

      end if

   - $e_{n,m} = \frac{\lambda_{n,m}^{wrong}}{\lambda_{n,m}^{corr} + \lambda_{n,m}^{wrong}}$;

   end for

   $m^+ = \arg\min_m (e_{n,m})$; //choose weak classifier with the lowest error

   $e_n = e_{n,m^+}$; $s_n^{sel} = h_{n,m^+}^{weak}$;

   if $e_n = 0$ or $e_n > \frac{1}{2}$ then exit

   end if

   $\alpha_n = \frac{1}{2}.\ln\left(\frac{1-e_n}{e_n}\right)$; //calculate voting weight

   // update importance weight

   if $s_n^{sel} = y$ **then** $\lambda = \lambda.\frac{1}{2.(1-e_n)}$;

   else $\lambda = \lambda.\frac{1}{2.e_n}$;

   end if

   $m^- = \arg\max_m (e_{n,m})$; $\lambda_{n,m^-}^{corr} = 1$; $\lambda_{n,m^-}^{wrong} = 1$;

   Get new h_{n,m^-}^{weak};

3. **End for**

---

first build a model by estimating the probability $P\left(1|f_j(x)\right) \sim N(\mu^+, \sigma^+)$ for posi-

tive labelled samples and $P\left(-1|f_j(x)\right) \sim N(\mu^-, \sigma^-)$ for negative labelled sam-

ples, where $f_j(x)$ evaluates this feature on the image x. The mean and variance are incrementally estimated by applying a Kalman filtering technique. Next, to estimate the hypothesis for Haar-Warelets, we use either simple threshold $h_j(x) = p_j .sign(f_j(x) - \theta_j)$ where $\theta_j = |\mu^+ + \mu^-|/2$, $p_j = sign(\mu^+ - \mu^-)$ or a Bayesian decision criterion:

$$h_j(x) = sign(P(1|f_j(x)) - P(-1|f_j(x))) \approx sign(g(f_j(x|\mu^+,\sigma^+) - g(f_j(x|\mu^-,\sigma^-))$$

where $g(x|\mu,\sigma)$ is a Gaussian probability density function. For histogram features (orientation histograms and LBPs), we use nearest neighbour learning D (e.g. Euclidean):

$$h_j(x) = sign(D(f_j(x), p_j) - D(f_j(x), n_j))$$

The cluster centers for positive p_j and negative n_j samples are learned by estimated the mean and the variance for each bin separately. All modules are based on the same type of classifier that is trained using the same features (For more details see [1]).

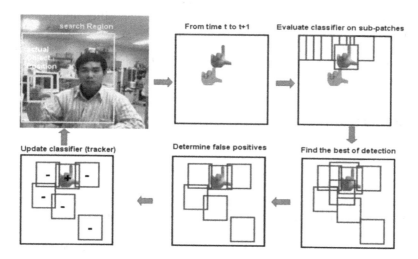

Fig. 1. Autonomous online learning – The main steps of the object's model learning are illustrated. Suitable updates for learning an object detector: Positive (red bounding box) and negatives (blue bounding boxes) as false positives are selected to update at time t.

3.4 Refinement of Generated Data

As we employ autonomous learning to train an object model, the classifier starts from only one initialized hand labelled training sample and performs update autonomously. At the first iteration, the classifier is updated once and then performs evaluation on current image to classify object and non-object classes. Since it is just updated only once, at this step, it has little knowledge about appearance of the object. So, it is rather "weak" in discriminating object class. Therefore, it may produce some wrongly

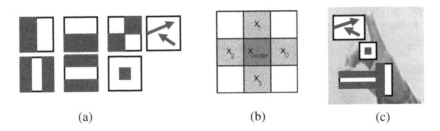

(a)   (b)   (c)

Fig. 2. Basic image features used. Haar-like features from Viola and Jones [3] and in addition orientation histograms (with 16 bins) from [19] and local binary patterns (LBP) [20] as features.

classified object sample, which is a "bad" sample to the database. The same situation may occur even after few iterations, i.e. the classifier has been updated just on few samples and does not cover various changing appearances of the object. Fortunately, because of our intelligent updating strategy, after sufficient number of iterations, the classifier becomes strong, which means the discriminative power has been increased, it can generate reliable samples to update itself and contribute to the database. What we are considering is the "bad" samples that have been added to the database at early stage of learning. The solution would be: to use the strong classifier, after training, to refine the generated training samples in the database. This can be easily done by applying the trained classifier on collected samples in the database. Bad samples, i.e. samples detected with low confident, will be removed. Thus, this results in a more "clean" data.

4 Experiments

In this section, we will demonstrate the capability of our proposed framework for the problem of data acquisition and training an object model from raw data. We conducted experiments on various objects with different complexities. In the following, we will present experiments and results mainly for the problem of human hand and face object learning for detection. But generic object types can also be applied. All experiments are set up as follows: each object model is represented by a classifier, which contains 150 weak classifiers, 50 selectors. First, we randomly initialize parameters for each classifier. Then a specific classifier is train for each object type with one click to initialize the object sample. Resulted video sequences are available on request.

4.1 Data sets

We have performed intensive experiments on several data sets with different complexities. The data sets are typically specific for different applications in computer vision. They include public available data and our recorded video. The goal of this section is to illustrate the effectiveness and robustness of our framework. First, we performed experiments on public available data sets to show our approach ability over very recent proposed approaches for tracking [21, 22]. The data sets include two typical challenging video sequences which we have selected from [22][1]. Second, to show the robustness of our framework, we perform experiments on several data sets to learn complex objects, include: deformed, articulated object such as hand movement recorded outdoor by a moving camera[2]; object in a context of very similar appearance, such as textured object on the same textured background. Third, further more, we show that our system is able to learn hand gestures model with hands of different persons. Data set for the third experiments includes video sequences which we recorded by a low resolution webcam camera. In the experiments we will show that our framework is able to: autonomous online learning for object's model from simple to complex objects quite well even in difficult circumstance, especially during the object's model learning the system also generates positive and negative samples for building data set.

* *Hand data* – video sequences: The first sequence shows a hand moving from the dark towards bright scene with rapid movements and postures changes, camera motion, changing of lighting conditions and arbitrary backgrounds. Other sequences contain hand movements with different gestures/postures in complex backgrounds.

* *Face data*: The sequence shows a person moving from dark to bright area with pose changes, illumination changes and clutter background.

* *Cars data*: The sequence shows cars in real scenario of out door environment [22].

4.2 Experiment Results

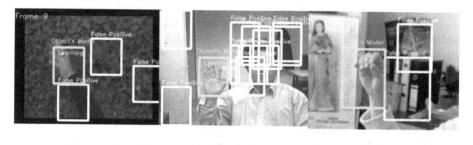

Fig. 3. Tracking results of autonomous object's model learning: true positive sample (yellow box), and false positive samples as negative (white box)

---

[1] http://www.cs.toronto.edu/~dross/ivt/
[2] http://www.movesinstitute.org/~kolsch/HandVu/VisionBasedHandTracking.wmv

Table 1. The result of object's model learning, number of labeled training samples have been collected (number of positives, negatives) and the detection rates of the test video sequences.

Data sets	Number of image frames have been annotated	Number of training samples	Number of training samples have been generated			Detection rate
			Number of Positive samples	Number of false Positive samples	Number of Negative samples extracted randomly from background	
Hand set 1	659	95	659	67	1343	100%
Hand set 2	600	67	600	190	1328	100%
Hand set 3	1977	129	1977	20	552	99.6%
Hand set 4	2130	135	2130	26	576	99.2%
Face set	462	26	462	126	1498	100%
Car set	659	164	659	183	1473	100%

5 Conclusion

In this paper, we have presented a new framework for generating visual object database meanwhile efficiently learning the object's model. Our system takes the advantages of the online boosting learning approach to build an autonomous learning object's model with reduction to minimal data acquisition effort. The system is composed of two main modules which are co-operated. The learning module, which is a discriminative classifier, learns efficient object representation online. The availability of the online classifier with intelligent update strategy in combining with tracking information allows to collect data samples during learning and to build a visual object database naturally. We have applied our framework for the problems of learning a detector for several object of interest, include: human hand, face and car models. The model learning module outperformed state-of-the-art online boosting learning approach in term of accuracy and stability. Database of each object have been generated efficiently, which contain informative, rich representation of considering object. Experiments have shown various capable applications of our proposed framework for visual data acquisition and object's model learning.

For future work, we plan to study more about the generalization ability of the autonomous learning algorithm. We also plan to use our result for further study of transfer learning. Moreover, multi tasks learning with automatic knowledge acquisition is a topic of interest.

References

1. Grabner, H., Bischof, H.: On-line boosting and vision. In: Proc. CVPR, IEEE, vol. 1, pp. 260–267 (2006)
2. Hertz, T., Bar-Hilled, A., Weinshall, D.: Learning distance functions for image retrieval. In: Proc. CVPR, IEEE, vol. 2, pp. 570–577 (2004)
3. Viola, P., Jones, M.: Rapid object detection using a boosted cascade of simple features. In: Proc. CVPR, IEEE, vol. I, pp. 511–518 (2001)

4. Tieu, K., Viola, P.: Boosting image retrieval. In: Proc. CVPR, IEEE, pp. 228–235 (2000)
5. Rowley, H., Baluja, S., Kanade, T.: Neural Network-based Face Detection. IEEE Trans. On PAMI 20(1), 23–38 (1998)
6. Platt, J.: Fast training of support vector machines using sequential minimal optimization. In: Advances in Kernel Methods - Support Vector Learning (1998)
7. Levin, A., Viola, P., Freund, Y.: Unsupervised improvement of visual detectors using co-training. In: Proc. IEEE CVPR, vol. I, pp. 626–633 (2003)
8. Nair, V., Clark, J.: An unsupervised, online learning framework for moving object detection. In: Proc. IEEE CVPR, vol. II, pp. 317–324 (2004)
9. Sung, K., Poggio, T.: Example-based learning for view-based face detection. IEEE Trans. on PAMI 20(1), 39–51 (1998)
10. Toyama, K., Krumm, J., Brumitt, B., Meyers, B.: Wallflower: Principles and Practice of Background Subtraction. In: Proc. of ICCV, pp. 255–261 (1999)
11. Elgamal, A., Harwood, D., Davis, L.: Non-parametric Model for Background Substraction. In: Proc. of ECCV (2000)
12. Sivic, J., Schaffalitzky, F., Zisserman, A.: Object level grouping for video shots. In: Proc. ECCV, vol. I, pp. 85–98 (2004)
13. Sivic, J., Everingham, M., Zisserman, A.: Person spotting: Video shot retrieval for face sets. In: Leow, W.-K., Lew, M., Chua, T.-S., Ma, W.-Y., Chaisorn, L., Bakker, E.M. (eds.) CIVR 2005. LNCS, vol. 3568, pp. 226–236. Springer, Heidelberg (2005)
14. Mikolajczyk, K., Schmid, C., Zisserman, A.: Human detection based on a probabilistic assembly of robust detectors. In: Pajdla, T., Matas, J(G.) (eds.) ECCV 2004. LNCS, vol. 3021, pp. 69–82. Springer, Heidelberg (2004)
15. Hewitt, R., Belongie, S.: Active learning in face recognition: Using tracking to build a face model. In: Proc. IEEE Workshop on Vision for Human-Computer Interaction (2006)
16. Wu, B., Nevatia, R.: Improving part based object detection by unsupervised, online boosting. In: Proc. IEEE Computer vision and Pattern Recognition (2007)
17. Javed, O., Ali, S., Shah, M.: Online detection and classification of moving objects using progressively improving detectors. In: Proc. IEEE CVPR (2005)
18. Oza, N.C., Russell, S.: Experimental comparisons of online and batch versions of bagging and boosting. In: Proc. ACM SIGKDD Intern. Conf. on Knowledge Discovery and Data Mining (2001)
19. Dalal, N., Triggs, B.: Histograms of oriented gradients for human detection. In: Proceedings, CVPR, San Diego, CA, USA, vol. 1, pp. 886–893 (2005)
20. Ojala, T., Pietikainen, M., Maenpaa, T.: Multiresolution gray-scale and rotation invariant texture classification with local binary patterns. Pattern Analysis and Machine Intelligence 24(7), 971–987 (2002)
21. Kolsch, M., Turk, M.: Fast 2D Hand Tracking with Flocks of Features and Multi-Cue Integration. In: IEEE Computer Society Conference on Computer Vision and Pattern Recognition Workshop, pp. 158–166 (2004)
22. Ross, D., Lim, J., Lin, R., Yang, M.H.: Incremental Learning for Robust Visual Tracking, the International Journal of Computer Vision. Special Issue: Learning for Vision (2007)

Finding the Most Interesting Association Rules by Aggregating Objective Interestingness Measures

Tri Thanh Nguyen Le[1], Hiep Xuan Huynh[1], and Fabrice Guillet[2]

[1] College of Information and Communication Technology
Can Tho University
lenguyenthanhtri@gmail.com
hxhiep@ctu.edu.vn
[2] Polytechnics School of Nantes University
fabrice.guillet@univ-nantes.fr

Abstract. Association rule post-processing is a research challenge in KDD. In this post-processing task, objective interestingness measures are very useful for finding interesting rules possessing certain characteristics. Till now, the usual method for using objective interestingness measures is to select one or several suitable measures for filtering rules. This paper proposes a new approach to aggregate a set of interestingness measures using the Choquet integral as an advanced aggregation operator. Since an objective interestingness measure is considered as a point of view on rule quality, the aggregation of a set of objective interestingness measures can extract rules satisfying many points of view. The experiment is carried out on different groups (i.e. different natures) of objective interestingness measures to observe their behaviors.

1 Introduction

Association mining [1][2] often generates a huge amount of rules, ranging from trivial to highly interesting. Thus, the post-processing step is crucial to assist the user in finding the rules which are truly beneficial. An approach toward rule post-processing is the use of objective interestingness measures [3][4][5][6][7][8] to assess their quality. Unlike their subjective counterparts, which require modeling the user's goal and knowledge, objective measures[1] rely only on the data, thus domain and user independent.

There exists an abundance of objective measures proposed in the literature (see [5] for a study about them). Each one is constructed based on a viewpoint on rule interestingness and possesses its own properties, making it suitable for some domains but not for others [4]. Some proposed ways of using objective measures are to select one or several suitable measures based on their properties [4][9] or in an interactive manner [6]. Since a measure can be considered as a criterion on rule interestingness, these methods consider different criteria separately. In other words, the best rules according to one measure may not necessarily be (and often are not) highly valued by others.

---

[1] In this paper, measure and interestingness measure, rule and association rule are used interchangeably.

D. Richards and B.-H. Kang (Eds.): PKAW 2008, LNAI 5465, pp. 40–49, 2009.

This paper proposes a new approach of filtering association rules by aggregating a set of objective measures to consider different points of view simultaneously. Different interestingness values given by different measures on a rule are aggregated together to produce a *global interestingness value* using the Choquet integral [10] - a sophisticated aggregation operator. The rules having the highest global value, called *the most interesting rules*, are often highly valued by all the measures used in the aggregation.

This paper is organized as follows. The next section presents the important concepts about the two areas: objective measures and the Choquet integral. The third section then discusses the application of the Choquet integral to the measure aggregation problem. The experimentation and conclusion take place in the remaining sections.

2 Interestingness Measures and Choquet Integral

2.1 Association Rules

Association rules [2] are implicative tendencies between two disjoint itemsets X and Y, denoted $X \to Y$, which means that if X occurs, Y is very likely to occur as well. Association rules are extracted from transaction databases in which each transaction contains a subset of items. A rule is associated with four cardinalities: the total number of transactions (n), the number of transactions containing X (n_X), the number of transactions containing Y (n_Y), and the number of transactions containing X but not Y ($n_{X\bar{Y}}$) (see Figure 1).

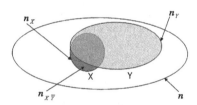

Fig. 1. The diagram of an association rule

2.2 Interestingness Measures

This work concerns only objective interestingness measures [11]. This kind of measure evaluates rules based solely on their cardinalities. More concretely, an objective interestingness measure is a function of four parameters $m(X \to Y) = f(n, n_X, n_Y, n_{X\bar{Y}})$. With plenty of objective measures in the literature, it is helpful to categorize them. In [8], these measures are classified according to their *subject* and *nature*. The subject of a measure is whether it has fixed value at independent point (where X and Y are stochastically independent: $n_X n_Y = n n_{XY}$) or equilibrium point (where the number of examples and counter examples are equal: $n_{XY} = n_{X\bar{Y}}$). The nature of a measure is whether it is descriptive or statistical. Descriptive measures are invariant with cardinality expansion (i.e. $m(n, n_X, n_Y, n_{X\bar{Y}}) = m(\alpha n, \alpha n_X, \alpha n_Y, \alpha n_{X\bar{Y}})$, with $\alpha > 0$). Otherwise, they are called statistical measures. The classification of popular measures according to these two aspects is given in [6].

2.3   Capacity Function

A capacity function (also called *fuzzy measure*) [12] on a set of measures $\mathcal{M}$ is a set function $\mu : 2^{\mathcal{M}} \to [0, 1]$ satisfying the following axioms:

- Normalized: $\mu(\emptyset) = 0$ and $\mu(\mathcal{M}) = 1$
- Monotonic: $\mu(A) \leq \mu(B)$ if $A \subseteq B \subseteq \mathcal{M}$

A capacity value of a set of measures is regarded as its weight or its importance. Capacity functions can be considered as an extension of probability vectors. Unlike probability vectors which assign a weight to each measure, capacity functions assign a weight for *each subset of measures* on condition that the weight of a set must not be smaller than that of any of its subsets. With a high number of parameters, capacity functions can model the interaction (or dependency) among measures. There are in general 3 kinds of interaction:

- *Negative interaction* (or complementarity): two measures m_i and m_j interact negatively together if their collective weight is less than the total of individual weights:

$$\mu(\{m_i, m_j\}) < \mu(m_i) + \mu(m_j)$$

  When two measures interact negatively, a rule appreciated by both measures is not much more interesting than a rule appreciated by only one of them.
- *Positive interaction* (or redundancy): reciprocally, if the collective weight of two measures m_i and m_j is greater than the total of individual weights, the interaction between them is positive:

$$\mu(\{m_i, m_j\}) > \mu(m_i) + \mu(m_j)$$

  When two measures interact positively, a rule highly valued by both measures is much more interesting than the one highly valued by only one measure.
- *Independence*: this is the case when no interaction exists between two measures m_i and m_j:

$$\mu(\{m_i, m_j\}) = \mu(m_i) + \mu(m_j)$$

  When every subset of measures is independent, the capacity is additive, i.e. $\mu(A \cup B) = \mu(A) + \mu(B)$, with $A \cap B = \emptyset$ and $A, B \subset \mathcal{M}$.

2.4   Choquet Integral

The Choquet integral [10] is used here as an aggregation operator. It is a generalization of the weighted mean by taking into account the interaction among measures. The Choquet integral C_μ with respect to a capacity function μ is defined as follows:

$$C_\mu(v_1, v_2, \ldots, v_k) = \sum_{i=1}^{k} v_{(i)} [\mu(A_{(i)}) - \mu(A_{(i+1)})] \tag{1}$$

with v_i is the interestingness value that the measure m_i gives to a rule and $v_{(1)}, v_{(2)}, \ldots, v_{(k)}$ are the permutation of $v_1, v_2, \ldots, v_k$ such that $v_{(1)} \leq v_{(2)} \leq \ldots \leq v_{(k)}$ and $A_{(i)} = \{m_{(i)}, m_{(i+1)}, \ldots, m_{(k)}\}$, $A_{(k+1)} = \emptyset$

The global value produced by the Choquet integral depends closely on the interaction between measures. Generally, a rule highly scored by complementary measures is assigned higher global interestingness value than a rule highly scored by redundant measures. When no interaction exists among measures, the capacity is additive and the Choquet integral degrades to the weighted mean.

2.5 Behavioral Analysis of the Choquet Integral

Given the large number of parameters in the Choquet integral, it is difficult to interpret its behaviors. To facilitate this task, some indices are proposed: the Shapley value [13] (or importance index) and the interaction index [14].

Shapley Value. In the Choquet integral context, the importance of a measure m_i is not only its weight $\mu(m_i)$ but also its contribution to every subset S of other measures $\mu(S \cup m_i) - \mu(S)$. The Shapley value [13] of a measure m_i, denoted $\phi_\mu(m_i)$, indicating its importance is defined as the mean of all these contributions:

$$\phi_\mu(m_i) = \sum_{T \subseteq \mathcal{M} \setminus m_i} \frac{(k - |T| - 1)! |T|!}{k!} [\mu(T \cup m_i) - \mu(T)] \qquad (2)$$

Interaction Index. To quantify the strength of interaction between two measures m_i and m_j, we consider the difference $(\Delta_{m_i m_j} \mu)(T)$ between the contribution of m_j in a subset of measures T when m_i is present and when m_i is absent:

$$(\Delta_{m_i m_j} \mu)(T) = [\mu(T \cup \{m_i, m_j\}) - \mu(T \cup m_i)] - [\mu(T \cup m_j) - \mu(T)]$$

with $T = \mathcal{M} \setminus \{m_i, m_j\}$. Intuitively, when m_i and m_j interact positively, the contribution of m_j is higher with the presence of m_i and reciprocally. The interaction index I_μ between two measures m_i and m_j is thus defined as [14]:

$$I_\mu(m_i, m_j) = \sum_{T \subseteq \mathcal{M} \setminus \{m_i, m_j\}} \frac{(n - |T| - 2)! |T|!}{(n - 1)!} (\Delta_{m_i m_j} \mu)(T) \qquad (3)$$

This value falls into the interval [-1,1]. It is positive (resp. negative) when the interaction between m_i and m_j is positive (resp. negative). The interaction value is zero when two measures are independent.

3 The Application of Choquet Integral to Measure Aggregation

3.1 The Measure Aggregation Problem

The aggregation problem consists in combining different (and sometime conflicting) points of view in order to form a global evaluation. Given a set of measures

$\mathcal{M} = \{m_1, m_2, \ldots, m_k\}$, which respectively give values $v_1, v_2, \ldots, v_k$ on a rule r based on its own point of view. The purpose is to find a "global interestingness value" v representing the overall evaluation of these measures on the rule r. The *most interesting rules* are the rules having the highest global interestingness value. These rules are often highly valued by all the measures used in the aggregation.

The simplest and most widely used form of aggregation is the *arithmetic mean* and the *weighted mean* (with weight of each criterion indicating its importance). This operator is not suitable for the aggregation of interestingness measures because it requires that the measures give values independently. Obviously, this condition is not satisfied. It is widely recognized that some measures are strongly dependent, i.e. they are strongly correlated on a rule set or even functionally dependent. These measures present very high redundancy between them. So, using weighted mean (and its variations) can lead to biased result because a group of highly similar measures can easily overwhelm other measures. Using the Choquet integral with an appropriate capacity function can avoid this problem.

3.2   The Construction of Capacity Function

The crucial part of using the Choquet integral is to model the interaction between measures via a capacity function. The usual ways of constructing capacity functions are to learn from a set of rules with their global interestingness value assigned by the user (see e.g [15] for a survey of these learning algorithms). In this work, we propose to construct a capacity function based on data. The purpose is to avoid the overwhelming effect induced by a group of highly similar measures. For this task, the more two measures are similar, the more negatively they interact with each other.

The Pearson coefficient is used here as a measure of similarity between two interestingness measures. Its value falls into the $[-1,1]$ interval. With ρ_{ij} the correlation value between two measures m_i and m_j, we have three cases:

- When $\rho_{ij} \gg 0$, m_i and m_j are strongly correlated. A rule which is given a high interestingness value by one measure is very likely to be considered interesting by another. These two measures thus present some redundancy. So, their interaction must be negative:

$$\mu(\{m_i, m_j\}) < \mu(m_i) + \mu(m_j)$$

  In the extreme case, when $\rho_{ij} = 1$, the two are substitutive, i.e. $\mu(\{m_i, m_j\}) = \max(\mu(m_i), \mu(m_j))$ (the maximum is to ensure the monotonicity of the capacity function, see Section 2.3).
- When $\rho_{ij} = 0$, m_i and m_j are stochastically independent, we have:

$$\mu(\{m_i, m_j\}) = \mu(m_i) + \mu(m_j).$$

- When $\rho_{ij} \ll 0$, m_i and m_j are anti-correlated. A rule which is highly valued by m_i is likely to be depreciated by m_j. It is unusual if a rule obtains high interestingness value on both measures. So, this rule should be appreciated. Thus, the interaction between them should be positive:

$$\mu(\{m_i, m_j\}) > \mu(m_i) + \mu(m_j).$$

We choose the simplest function passing these reference points, which is a straight line having the function (see Figure 2):

$$\mu(\{m_i, m_j\}) = \max[\mu(m_i), \mu(m_j)] + (1 - \rho) \min[\mu(m_i), \mu(m_j)] \qquad (4)$$

Generalizing the above formula for a set of measures S :

$$\mu(S) = \max_{m_i \in S} \mu(S \backslash \{m_i\}) + (1 - \bar{\rho}) \mu(m_i) \qquad (5)$$

with $\bar{\rho}$ denotes the mean of correlation coefficient between the measure m_i with all the other measures in S.

The value of singletons (i.e. $\mu(m_i)$) can be assigned by the user. Higher value on the measure m_i means higher importance placed on this measure. Without a priori preference on the measures, they can be assigned equal weights.

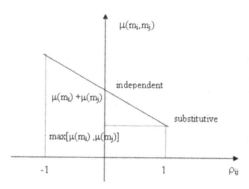

Fig. 2. The dependency of capacity value with the correlation value between two measures

4   Experimentation

In this section, the experiments are carried out on two rule sets of contrasting nature: the correlated rule set MUSHROOM and the weakly correlated rule set T5I2D10K. The MUSHROOM data set, taken from UC Irvine Machine-Learning Database Repository [16], contains 23 attributes corresponding to the species of gilled mushrooms. The T5I2D10K is a synthetic data set obtained by simulating the customer transactions with the average transaction length is 5 (T5), the average itemset size is 2 (I2) and the number of item is 100 (D10k). The two rule sets extracted by Apriori algorithm [2] have over one hundred thousand rules.

The aggregation is realized on two groups of measures: measures of deviation from equilibrium and measures of deviation from independence (see section 2.2). Because of commensurability handling, only measures having finite co-domain are chosen for aggregation. In each aggregation result, the behaviors of the Choquet integral as well as the characteristics of the most interesting rules are analyzed in order to draw comparison between them.

4.1   Aggregation of Measures of Deviation from Equilibrium

The measures chosen from this cluster are Confidence (CONF), Descriptive Confirm (DesCF), Descriptive Confirmed Confidence (DesCONF), Laplace (LAP) and IPEE.

Interaction Index. The interaction values relate closely to the correlation coefficient between measures (see Table 1 and 2). The more correlated are two measures, the more redundant they are and the more negative the interaction between them is. The correlation between measures is very different on two rule sets (because of their contrasting nature). The pairs (Confidence, Descriptive Confirm) and (Descriptive Confirm, Descriptive Confirm Confidence) are more strongly correlated on MUSHROOM rule set than on T5I2D10K, while IPEE is more loosely correlated with other measures. So, the two former pairs interact more negatively on MUSHROOM than on T5I2D10K and IPEE interact less negatively with other measures. Despite the difference of other measures on two rule sets, the group (Confidence, Descriptive Confirmed Confidence, Laplace) always correlates strongly with each other and possess the most negative interaction on both rule sets.

Table 1. The correlation coefficient and interaction value between 5 measures of deviation from equilibrium on MUSHROOM rule set

Correlation coefficient				
	CONF	DesCF	DesCONF	IPEE
CONF				
DesCF	0.88			
DesCONF	1.0	0.88		
IPEE	0.87	0.79	0.87	
LAP	1.0	0.88	1.0	0.87

Interaction value				
	CONF	DesCF	DesCONF	IPEE
CONF				
DesCF	-0.136			
DesCONF	-0.184	-0.136		
IPEE	-0.130	-0.109	-0.130	
LAP	-0.184	-0.136	-0.184	-0.130

Table 2. The correlation coefficient and interaction value between 5 measures of deviation from equilibrium on T5I2D10K rule set

Correlation coefficient				
	CONF	DesCF	DesCONF	IPEE
CONF				
DesCF	0.54			
DesCONF	1.0	0.54		
IPEE	0.94	0.39	0.94	
LAP	1.0	0.56	1.0	0.94

Interaction value				
	CONF	DesCF	DesCONF	IPEE
CONF				
DesCF	-0.007			
DesCONF	-0.137	-0.007		
IPEE	-0.121	0.052	-0.121	
LAP	-0.136	-0.012	-0.136	-0.118

Shapley Value. Since the initial weights of all measures are the same, the importance of a measure depends on its interactions with other measures. The less negatively its interactions are, the more important it is. Hence, Confidence, Descriptive Confirmed Confidence, Laplace possess the lowest importance on both rule sets. On MUSHROOM rule set, Descritive Confirm possesses higher importance than T5I2D10K while IPEE possesses lower importance (because of their interaction discussed above).

The Most Interesting Rules. The 10 most interesting rules (i.e. rules having the highest global interestingness values) are very similar together (see Table 3). They tend to

Table 3. The 10 most interesting rules obtained by aggregation of 5 measures of deviation from equilibrium

n^o	n	n_X	n_Y	$n_{X\bar{Y}}$	Rule
Mushroom rule set					
1	8416	8200	8216	8	FREE → v_col=WHITE
2	8416	8216	8200	24	v_col=WHITE → FREE
3	8416	7568	8200	0	ONE v_col=WHITE → FREE
4	8416	7576	8216	8	FREE ONE → v_col=WHITE
5	8416	7768	8200	192	ONE → FREE
6	8416	7768	8192	200	ONE → FREE v_col=WHITE
7	8416	7768	8216	200	ONE → v_col=WHITE
8	8416	6608	8216	0	CLOSE FREE → v_col=WHITE
9	8416	6632	8200	24	CLOSE FREE ONE → FREE
10	8416	6272	8216	0	CLOSE FREE ONE → v_col=WHITE
T5I2D10K rule set					
1	9650	971	1720	75	77 → 1
2	9650	565	1878	25	41 88 → 71
3	9650	570	1122	30	71 88 → 41
4	9650	452	1601	14	0 67 → 31
5	9650	487	1601	23	67 → 31
6	9650	432	1601	13	0 57 67 → 31
7	9650	431	1601	13	0 12 67 → 31
8	9650	487	1124	24	67 → 12
9	9650	434	1601	15	0 12 57 → 31
10	9650	487	784	25	67 → 57

Table 4. The 10 most interesting rules obtained by aggregation of 10 descriptive measures of deviation from independence

n^o	n	n_X	n_Y	$n_{X\bar{Y}}$	Rule
Mushroom rule set					
1	8416	1840	1848	8	EVAN NAR → SEVERAL sp_col=WHITE
2	8416	1848	1840	16	SEVERAL sp_col=WHITE → EVAN NAR
3	8416	1768	1808	8	EVAN POIS → ? NAR
4	8416	1808	1768	48	? NAR → EVAN POIS
5	8416	1768	1848	8	EVAN POIS → SEVERAL sp_col=WHITE
6	8416	1848	1768	88	SEVERAL sp_col=WHITE → EVAN POIS
7	8416	1768	1880	8	EVAN POIS → ? SEVERAL
8	8416	1808	1824	48	? NAR → POIS sp_col=WHITE
9	8416	1824	1808	64	POIS sp_col=WHITE → ? NAR
10	8416	1808	1832	48	? NAR → EVAN SEVERAL
T5I2D10K rule set					
1	9650	121	130	4	12 55 69 97 → 44 73
2	9650	121	130	4	12 55 73 97 → 44 69
3	9650	126	130	6	12 55 97 → 44 69
4	9650	126	130	6	12 55 97 → 44 73
5	9650	125	132	5	12 69 97 → 44 55
6	9650	125	130	6	12 69 97 → 44 73
7	9650	130	126	10	44 69 → 12 55 97
8	9650	130	126	10	44 73 → 12 55 97
9	9650	126	132	6	12 73 97 → 44 55
10	9650	125	126	8	12 69 97 → 44 55 73

possess low counter examples (which satisfies the view point of Confidence, Descriptive Confirmed Confidence and Laplace), very high antecedent and consequent cardinality (which conforms to the view point of Descriptive Confirm).

4.2   Aggregation of Descriptive Measures of Deviation from Independence

This group consists of 10 measures: Dependency, Gini Index, J-measure, Kappa, Mutual Information, Pavillon, Phi-coefficient, TIC, Yule's Q and Yule's Y.

Interaction Index. The measures are more strongly correlated on MUSHROOM rule set than on T5I2D10K. In fact, on the former rule set, all measures are strongly correlated, except TIC. Thus, the interactions between TIC and other measures are weakly negative. The strongest interaction exists between the pair (Dependency, Pavillon) and (Yule's Q, Yule's Y). On the later rule set, there are four groups exhibiting strong negative interaction (Dependency, Pavillon), (Kappa, Phi-coefficient, Mutual Information), (Gini Index, J-measure) and (Yule's Q, Yule's Y).

Shapley Value. On both rule sets, TIC is assigned the highest importance value because it possesses low redundancy with other measures. On MUSHROOM, Yule's Q, Yule's Y and Mutual Information possess higher importance than on T5I2D10K while Dependency, Gini-Index, J-measure and Pavillon are assigned lower importance (see Figure 4).

The Most Interesting Rules. The most interesting rules resulting from the aggregation are very similar together on both rule set. They all possess very low counter examples as well as low cardinalities of antecedent and consequent (Table 4). Rules of T5I2D10K tend to possess lower counter examples because Dependency and Pavillon possess higher importance. These measures prefer rules with low counter examples.

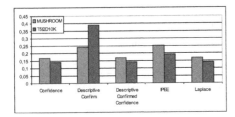
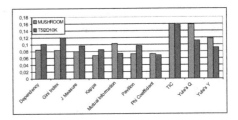

Fig. 3. Shapley value of measures of deviation from equilibrium

Fig. 4. Shapley value of descriptive measures of deviation from independence

4.3 Remarks

Through the experiments carried out on different rule sets of contrasting natures, it is observed that the behaviors of the Choquet integral change with the rule set used because the correlation coefficient between measures is context dependent. Despite these differences, the characteristics of most interesting rules remain quite similar. This reflects the point of view of measures in each group.

5   Conclusion

This paper proposes a new method for using objective measures to search for interesting rules in a large rule set. While other methods proposed in the literature focus on choosing one or several measures which best suit the applied domain and treat these measures separately, our approach aims to aggregate various measures in order to find rules that satisfy numerous points of view simultaneously. This direction flows naturally from the proliferation of proposed measures in the literature. In our work, the Choquet integral, a suitable aggregation operator in this context, is chosen and a method of constructing the capacity based on data is proposed to take into account the dependence between measures. Experimentation results verified the behaviors of the Choquet integral and the characteristics of the most interesting rules.

The measure aggregation alleviates the burden of measure selection. But the user still has to consider which measures to use because unsuitable measures may bias the aggregation result. The implementation of this aggregation method is already incorporated in ARQAT framework [6]. Further researches in this direction consist in finding a more sophisticated measure of similarity between interestingness measures to replace the linear correlation. The construction of capacity based on user's opinion instead of data is also an interesting approach.

References

1. Agrawal, R., Imielinski, T., Swami, A.: Mining association rules between sets of items in large database. In: Proceedings of 1993 ACM SIGMOD International Conference on Management of Data, pp. 207–216 (1993)
2. Agrawal, R., Srikant, R.: Fast algorithm for mining association rules. In: Proceedings of the 20th Very Large Databases conference, pp. 487–499 (1994)

3. Klemettinen, M., Mannila, H., Ronkainen, P., Toivonen, H., Verkamo, A.I.: Finding interesting rules from large sets of discovered association rules. In: Proceedings of the Third International Conference on Information and Knowledge Management (CIKM 1994), pp. 401–407 (1994)
4. Tan, P., Kumar, V., Srivastava, J.: Selecting the right interestingness measure for association patterns. Information Systems 4(29), 293–313 (2004)
5. Guillet, F.: Mesures de la qualité des connaissances en ECD. In: EGC 2004 Actes des Tutoriels, 4e Conférence Francophone en Extraction et Gestion des Connaissances, pp. 1–60 (2004) (in French)
6. Huynh, H.X., Guillet, F., Blanchard, J., Kuntz, P., Gras, R., Briand, H.: A graph-based clustering approach to evaluate interestingness measures: a tool and a comparative study. In: Quality Measures in Data Mining. Studies in Computational Intelligence, pp. 25–50. Springer, Heidelberg (2007)
7. Blanchard, J., Guillet, F., Gras, R., Briand, H.: Using information-theoretic measures to assess association rules interestingness. In: ICDM 2005, Proceedings of the 5th IEEE International Conference on Data Mining, pp. 191–200 (2005)
8. Blanchard, J., Guillet, F., Gras, R., Briand, H.: Assessing rule interestingness with a probabilistic measure of deviation from equilibrium. In: ASMDA 2005, Proceeding of the 11th International Symposium on Applied Stochastic Models and Data Analysis, pp. 191–200 (2005)
9. Geng, L., Hamilton, H.J.: Interestingness measures for data mining: A survey. ACM Computing Survey 38(3), 1–32 (2006)
10. Choquet, G.: Theory of capacities. Annales de l'Institut Fourier 5, 131–295 (1953)
11. Silberchatz, A., Tuzhilin, A.: What makes pattern interesting in knowledge discovery systems. IEEE Transactions on Knowledge and Data Engineering 5, 970–974 (1996)
12. Sugeno, M.: Theory of fuzzy integrals and its applications. Ph.D thesis, Tokyo Institute of Technology (1974)
13. Shapley, L.: A value for n-person games. In: Kuhn, H., Tucker, A. (eds.) Contributions to the Theory of Games. Annals of mathematics studies edn., vol. 2, pp. 307–317. Princeton University Press, Princeton (1953)
14. Murofushi, T., Soneda, S.: Techniques for reading fuzzy measures (iii): interaction index. In: The 9th Fuzzy System Symposium, pp. 693–696 (1993)
15. Grabisch, M., Kojadinovic, I., Meyer, P.: A review of capacity identification methods for Choquet integral based multiattribute utility theory: Applications of the KAPPALAB R Package. European Journal of Operational Research 186(2), 766–785 (2007)
16. Asuncion, A., Newman, D.: UCI machine learing repository. University of California, Irvine, School of Information and Computer Sciences (2007),
   http://www.ics.uci.edu/~mlearn/MLRepository.html

Pruning Strategies Based on the Upper Bound of Information Gain for Discriminative Subgraph Mining

Kouzou Ohara, Masahiro Hara, Kiyoto Takabayashi,
Hiroshi Motoda, and Takashi Washio

The Institute of Scientific and Industrial Research, Osaka University,
8-1, Mihogaoka, Ibaraki, Osaka 567-0047, Japan
{ohara,hara,kiyoto_ra,motoda,washio}@ar.sanken.osaka-u.ac.jp

Abstract. Given a set of graphs with class labels, discriminative sub-graphs appearing therein are useful to construct a classification model. A graph mining technique called Chunkingless Graph-Based Induction (Cl-GBI) can find such discriminative subgraphs from graph structured data. But, it sometimes happens that Cl-GBI cannot extract subgraphs that are good enough to characterize the given data due to its time and space complexities. Thus, to improve its efficiency, we propose pruning methods based on the upper-bound of information gain that is used as a criterion for discriminability of subgraphs in Cl-GBI. The upper-bound of information gain of a subgraph is the maximal one that its super graph can achieve. By comparing the upper-bound of each subgraph with the best information gain at the moment, Cl-GBI can exclude unfruitful sub-graphs from its search space. Furthermore, we experimentally evaluate the effectiveness of the pruning methods on a real world and artificial datasets.

Keywords: Graph mining, information gain, discriminative subgraph, data mining.

1 Introduction

Over the last decade, there has been much research work on data mining which intends to find useful and interesting knowledge from massive data. A number of algorithms have been proposed in recent years especially on graph mining, *i.e.*, mining frequent subgraphs from graph structured data because of the high expressive power of graph representation[1,2,3,4,5,6]. Since subgraphs found by these algorithms characterize the given graph dataset, it is natural to consider classification of graphs as one of practical applications of graph mining. In that case, as the index to find useful subgraphs, the discriminativity of subgraphs becomes more important. In other words, discriminative subgraphs appearing in a given set of graphs with class labels are useful to construct a classification model.

Chunkingless Graph Based Induction (Cl-GBI) [7] is one of the latest algorithms in graph mining, and can find such discriminative subgraphs from graph

D. Richards and B.-H. Kang (Eds.): PKAW 2008, LNAI 5465, pp. 50–60, 2009.
© Springer-Verlag Berlin Heidelberg 2009

structured data with class labels by taking into account not only the frequency of subgraphs but also a measure to evaluate the discriminativity such as information gain [8]. Cl-GBI is an extension of Graph Based Induction (GBI) [2] and similarly to GBI and its another extension, Beam-wise GBI(B-GBI) [3], extracts typical subgraphs from graph-structured data on the basis of the stepwise pair expansion principle which recursively chunks two adjoining nodes. But, Cl-GBI actually never chunks them. Instead, it regards a pair of adjoining nodes as a *pseudo-node* and assigns a new label to it, which is called *pseudo-chunking*. Introduction of pseudo-chunking can fully solve the reported problems caused by chunking, *i.e.*, ambiguity in selecting nodes to chunk and incompleteness of the search due to overlooking some overlapping subgraphs. This is because every original node is available to make a new pseudo-node at any time in Cl-GBI. However Cl-GBI requires more time and space complexities in exchange of gaining the ability of extracting overlapping subgraphs. Thus, it sometimes happens that Cl-GBI cannot extract subgraphs that need be large enough to describe characteristics of data within a limited time and a given computational resource. In such a case, extracted subgraphs may not be so much of useful for classifying the given graphs.

From this background, in this paper, we consider the problem of finding discriminative subgraphs from graph structured data with class labels by using Cl-GBI, and to reduce its computational cost, propose pruning strategies which can exclude unfruitful pseudo-nodes (subgraphs) or node pairs from its search space. For this purpose, we focus on the fact that given a subgraph g the upper-bound of information gain achieved by its super-graphs is computable, and introduce two kinds of pruning strategies based on the upper-bound into Cl-GBI. Furthermore, we implement Cl-GBI involving the pruning steps and apply it to both synthetic and real world datasets. Through those experiments, we show the effectiveness of the proposed pruning strategies.

2   Finding Discriminative Subgraphs with Chunkingless Graph-Based Induction(Cl-GBI)

2.1   Overview of Cl-GBI

Cl-GBI is an extension of Graph Based Induction (GBI) [2] that can extract frequent subgraphs from graph structured data by stepwise pair expansion. Stepwise pair expansion is an essential operation in GBI, which recursively generates new nodes by selecting pairs of two adjoining nodes according to a certain criterion such as frequency and replacing all of their occurrences in graphs with a node having a newly assigned label. Namely each graph is rewritten each time a pair is chunked, and never restored in any subsequent chunking.[1] Although thanks to this chunking mechanism, GBI can efficiently extract subgraphs from either a huge single graph or a set of graphs, it involves ambiguity in selecting

---

[1] This does not mean that the link information of the original graphs is lost. It is always possible to restore how each node is connected in the extracted subgraphs.

Input. A graph database G, a beam width b, the maximal number of levels L, a criterion for ranking pairs to be pseudo-chunked C, a necessary condition that resulting subgraphs must satisfy θ;

Output. A set of typical subgraphs S;

Step 1. Extract all the pairs consisting of two connected nodes in G, register their positions using node id (identifier) sets (from the 2nd level on, extract all the pairs consisting of two connected nodes with at least one node being a new pseudo-node).

Step 2. Count frequencies of extracted pairs and eliminate those which do not satisfy the necessary condition θ.

Step 3. Select the best b pairs according to C from among the remaining pairs at Step 2 (from the 2nd level on, from among the unselected pairs in the previous levels and the newly extracted pairs). Each of the b selected pairs is registered as a new pseudo-node. If either or both nodes of the selected pair are not original but pseudo-nodes, they are restored to the original subgraphs before registration.

Step 4. Assign a new label to each pair selected at Step 3 but do not rewrite the graphs. Go back to Step 1.

Fig. 1. Algorithm of Cl-GBI

nodes to chunk, which causes a crucial problem, *i.e.*, possibility of overlooking some overlapping subgraphs due to inappropriate chunking order. Beam search adopted by Beam-wise GBI(B-GBI) [3] can alleviate this problem by chunking the best b (beam width) pairs w.r.t. a given criterion and copying each graph into respective states, but not completely solve it because chunking process is still involved.

In contrast to GBI and B-GBI, Cl-GBI does not chunk a selected pair, but regards it as a *pseudo-node* and assigns a new label to it. Thus, graphs are not "compressed" nor copied over the iterative *pseudo-chunking* process. We refer to each iteration in Cl-GBI as "level". The algorithm of Cl-GBI is shown in Fig. 1. The search of Cl-GBI is controlled by the following parameters: a necessary condition θ to be satisfied by resulting subgraphs, the number of pairs pseudo-chunked at once (beam width) b, a criterion C to rank node pairs for selecting the best b pairs, and the maximal number of levels L. In other words, at each level, the best b node pairs w.r.t. C are selected from among those which satisfy the condition θ, and pseudo-chunked. On one hand, in case of finding frequent subgraphs, the minimum frequency threshold is set to θ and the frequency of subgraphs is adopted as the criterion C. On the other hand, when finding discriminative subgraphs, another criterion such as the information gain of subgraphs could be used as C.

Although Cl-GBI completely alleviated the problem caused by the original chunking as mentioned above, introducing the pseudo-node considerably increased the number of pairs extracted at Step 1 of the algorithm shown in Fig. 1. As a result, Cl-GBI requires more time and space complexities in exchange of gaining the ability of extracting overlapping patterns. Thus, it sometimes happens that Cl-GBI cannot extract subgraphs that need be large enough to describe characteristics of data within a limited time and a given computational resource. In such a case, extracted subgraphs may not be so much of useful for the users.

2.2 Definition of Information Gain for Finding Discriminative Subgraphs

In this paper, we deal with the problem of finding discriminative subgraphs with Cl-GBI from a set of graphs G, in which each graph has a class label. To simplify the discussion, in what follows we suppose that there are only two classes and refer to them as "positive" and "negative" for convenience, which are denoted by "+" and "−", respectively in equations. But, the discussion in this paper can be easily extended to the case that there are more than two classes. In addition, we adopt information gain as a typical criterion to evaluate the discriminativity of subgraphs.

Suppose that a subgraph g is given and that G is divided into two subsets G_g and $G_{\bar{g}}$, which consist of graphs including and not including g, respectively. Then, the information gain of g is given as the reduction in entropy, $i.e.$, uncertainty of the class assignment to sets of graphs by majority voting between before and after dividing G. Namely, the higher the information gain is, the purer the class distributions in G_g and $G_{\bar{g}}$. Formally, the information gain of g, $Gain(g, G)$, is defined as follows:

$$Gain(g, G) = Ent(G) - \sum_{i \in \{g, \bar{g}\}} \frac{|G_i|}{|G|} Ent(G_i), \tag{1}$$

where $Ent(G)$ and $Ent(G_i)$ $(i \in \{g, \bar{g}\})$ are entropies of G and G_i, respectively, and defined as follows:

$$Ent(G) = - \sum_{j \in \{+, -\}} \frac{|G_j|}{|G|} \log_2 \frac{|G_j|}{|G|}, \tag{2}$$

$$Ent(G_i) = - \sum_{j \in \{+, -\}} \frac{|G_{i_j}|}{|G_i|} \log_2 \frac{|G_{i_j}|}{|G_i|}, \tag{3}$$

where G_+ and G_- are disjoint subsets of G such that $G_+ \cup G_- = G$ and consist of graphs which belong to the class positive and those which belong to the class negative, respectively; G_{g+}, G_{g-}, $G_{\bar{g}+}$, and $G_{\bar{g}-}$ are defined as $G_g \cap G_+$, $G_g \cap G_-$, $G_{\bar{g}} \cap G_+$, and $G_{\bar{g}} \cap G_-$, respectively. Since a node pair represents a subgraph in Cl-GBI, after counting the frequency of an extracted pair g at Step 2 of the algorithm shown in Fig. 1, all of $|G_g|$, $|G_{g+}|$, $|G_{g-}|$, $|G_{\bar{g}+}|$, and $|G_{\bar{g}-}|$ are available.

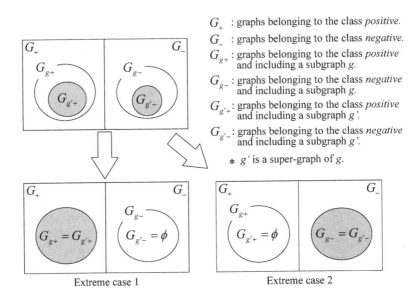

G_+ : graphs belonging to the class *positive*.

G_- : graphs belonging to the class *negative*.

G_{g+} : graphs belonging to the class *positive* and including a subgraph g.

G_{g-} : graphs belonging to the class *negative* and including a subgraph g.

$G_{g'+}$: graphs belonging to the class *positive* and including a subgraph g'.

$G_{g'-}$: graphs belonging to the class *negative* and including a subgraph g'.

∗ g' is a super-graph of g.

Fig. 2. Two extreme cases caused by a super-graph of g

3   Pruning Strategies Based on the Upper-Bound of Information Gain

As mentioned above, Cl-GBI yields much more node pairs (subgraphs) than GBI or B-GBI. If we know that for an extracted pair, none of its super-graphs can give the information gain which is greater than the maximum one at the moment, we could discard such an unfruitful pair and reduce the computational cost of subsequent levels. Fortunately, since the information gain is a convex function, it is theoretically shown that the upper-bound of the information gain is computable[9].

Let g' be a super-graph of a subgraph g. Note that $G_{g'_+} \cap G_{g'_-} = \emptyset$ and $G_{g'_+} \cup G_{g'_-} = G_{g'}$. In addition, since g' is a super-graph of g, $|G_{g'_+}| \leq |G_{g+}|$ and $|G_{g'_-}| \leq |G_{g-}|$. Then, if either "$|G_{g'_+}| = |G_{g+}|$ and $|G_{g'_-}| = 0$" or "$|G_{g'_-}| = |G_{g-}|$ and $|G_{g'_+}| = 0$" holds, $Gain(g', G)$ takes the maximum value, which is the upper-bound of the information gain derived from the super-graphs of g. Figure 2 illustrates these extreme cases. Thus, according to Equation (1), the upper-bound of information gain $u(g, G)$ is defined as follows:

$$u(g, G) = \max(u^+(g, G), u^-(g, G)), \tag{4}$$

where $u^+(g, G)$ and $u^-(g, G)$ are defined as follows:

$$u^+(g, G) = Ent(G) - \left(\frac{|G_{g+}|}{|G|} \left(\frac{|G_{g+}|}{|G_{g+}|} \log_2 \frac{|G_{g+}|}{|G_{g+}|} + 0 \right) + \frac{|G - G_{g+}|}{|G|} Ent(G - G_{g+}) \right)$$

$$= Ent(G) - \frac{|G - G_{g+}|}{|G|} Ent(G - G_{g+}), \tag{5}$$

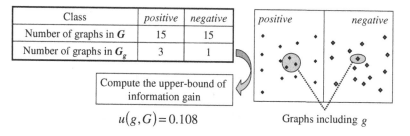

Class	positive	negative
Number of graphs in G	15	15
Number of graphs in G_g	3	1

Compute the upper-bound of information gain

$u(g, G) = 0.108$

positive | negative

Graphs including g

Fig. 3. An Example of post-pruning

$$u^-(g,G) = Ent(G) - \left(\frac{|G_{g_-}|}{|G|} \left(0 + \frac{|G_{g_-}|}{|G_{g_-}|} \log_2 \frac{|G_{g_-}|}{|G_{g_-}|} \right) + \frac{|G - G_{g_-}|}{|G|} Ent(G - G_{g_-}) \right)$$

$$= Ent(G) - \frac{|G - G_{g_-}|}{|G|} Ent(G - G_{g_-}). \tag{6}$$

The following lemma is directly derived from the results shown in [9].

Lemma 1. *For a subgraph g appearing in G, there is no super-graph of g that can achieve the information gain greater than $u(g, G)$.*

Thanks to this lemma, after the frequency counting step, we can find and safely prune unfruitful pairs by comparing the upper-bound of information gain derived from each newly extracted pair with the maximum information gain at the moment, τ. We call this pruning the *post-pruning*, which is formally defined as follows:

Definition 1. (Post-pruning)
Discard a node pair (subgraph) g if $u(g, G) < \tau$.

It is expected that the post-pruning can reduce the computational cost to make pairs in subsequent levels and save the memory space to register node pairs. For example, in Fig.3, since $|G_{g_+}| = 3$ and $|G_{g_-}| = 1$, according to the above definition, one can obtain $u(g, G) = 0.108$. Then, if $\tau > 0.108$, g is discarded.

In addition to this naive pruning, we consider another one which can be inserted before the frequency counting of Cl-GBI because it is one of time consuming tasks in Cl-GBI. If we can prune some node pairs before frequency counting, it could reduce its computational cost. However, unlike the post-pruning, we cannot directly compute the upper-bound of information gain derived from a node pair g because we have not known $|G_g|$, $|G_{g_+}|$, $|G_{g_-}|$, $|G_{\bar{g}_+}|$, and $|G_{\bar{g}_-}|$ yet at the moment. Instead, we know the frequency of each node composing the pair because it has been computed at the previous level.

Consequently, we adopt the most optimistic estimation $\hat{u}(g, G)$ instead of the actual $u(g, G)$. Let g_1 and g_2 be nodes composing the node pair g. Then, $|G_{g_+}|$ is at most $|G_{g_{1_+}} \cap G_{g_{2_+}}|$. As well, $|G_{g_-}|$ is at most $|G_{g_{1_-}} \cap G_{g_{2_-}}|$. Thus, $u(g, G)$ is maximized either when $|G_{g_+}| = |G_{g_{1_+}} \cap G_{g_{2_+}}|$ and $|G_{g_-}| = 0$ ($G_{g_-} = \emptyset$) or when $|G_{g_+}| = 0$ ($G_{g_+} = \emptyset$) and $|G_{g_-}| = |G_{g_{1_-}} \cap G_{g_{2_-}}|$ as shown in Fig. 4. We used

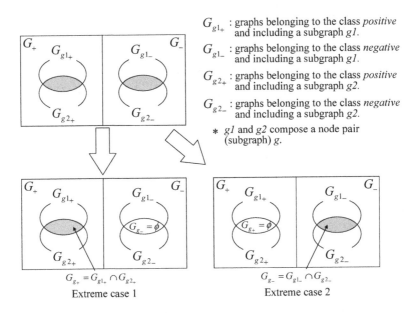

Fig. 4. Two extreme cases caused by a node pair (subgraph) g consisting of nodes g_1 and g_2

this maximally estimated upper-bound as $\hat{u}(g, G)$, which is formally defined as follows:

$$\hat{u}(g, G) = \max(\hat{u}^+(g, G), \hat{u}^-(g, G)), \tag{7}$$

where $\hat{u}^+(g, G)$ and $\hat{u}^-(g, G)$ are defined as follows:

$$\hat{u}^+(g, G) = Ent(G) - \left(\frac{|G_{\hat{g}+}|}{|G|} \left(\frac{|G_{\hat{g}+}|}{|G_{\hat{g}+}|} \log_2 \frac{|G_{\hat{g}+}|}{|G_{\hat{g}+}|} + 0 \right) + \frac{|G - G_{\hat{g}+}|}{|G|} Ent(G - G_{\hat{g}+}) \right)$$

$$= Ent(G) - \frac{|G - G_{\hat{g}+}|}{|G|} Ent(G - G_{\hat{g}+}), \tag{8}$$

$$\hat{u}^-(g, G) = Ent(G) - \left(\frac{|G_{\hat{g}-}|}{|G|} \left(0 + \frac{|G_{\hat{g}-}|}{|G_{\hat{g}-}|} \log_2 \frac{|G_{\hat{g}-}|}{|G_{\hat{g}-}|} \right) + \frac{|G - G_{\hat{g}-}|}{|G|} Ent(G - G_{\hat{g}-}) \right)$$

$$= Ent(G) - \frac{|G - G_{\hat{g}-}|}{|G|} Ent(G - G_{\hat{g}-}), \tag{9}$$

where $G_{\hat{g}+} = G_{g1+} \cap G_{g2+}$ and $G_{\hat{g}-} = G_{g1-} \cap G_{g2-}$.

We call the pruning step based on $\hat{u}(g, G)$ the *pre-pruning*, which is formally defined as follows:

Definition 2. (Pre-pruning)
Discard a subgraph g derived from a pair of nodes if $\hat{u}(g, G) < \tau$.

For example, Fig.5 shows the case that $G_{\hat{g}+} = 4$ and $G_{\hat{g}-} = 3$. In this case, according to the above definition, $\hat{u}(g, G) = 0.148$. If $\tau > 0.148$, then g is discarded because the information gain of any super graph of g cannot be greater than τ.

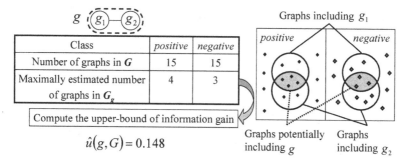

Fig. 5. An Example of pre-pruning

Input. A graph database G, a beam width b, the maximal number of levels L, a criterion for ranking pairs to be pseudo-chunked C, a necessary condition that resulting subgraphs must satisfy θ, a threshold of the upper-bound of information gain τ;

Output. A set of typical subgraphs S;

Step 1. Extract all the pairs consisting of two connected nodes in G, register their positions using node id (identifier) sets (from the 2nd level on, extract all the pairs consisting of two connected nodes with at least one node being a new pseudo-node);

Step 2.1. (Pre-pruning) For each extracted pair g, calculate $\hat{u}(g,G)$ and discard g if $\hat{u}(g,G) < \tau$;

Step 2.2. For each remaining pair, count its frequency, discard it if it does not satisfy the necessary condition θ, calculate its information gain, and update τ if its information gain is greater than τ;

Step 2.3. (Post-pruning) For each remaining pair g, calculate $u(g,G)$ and discard g if $u(g,G) < \tau$;

Step 3. Select the best b pairs according to C from among the remaining pairs at Step 2 (from the 2nd level on, from among the unselected pairs in the previous levels and the newly extracted pairs). Each of the b selected pairs is registered as a new node. If either or both nodes of the selected pair are not original but pseudo-nodes, they are restored to the original subgraphs before registration.

Step 4. Assign a new label to each pair selected at Step 3 but do not rewrite the graphs. Go back to Step 1.

Fig. 6. Algorithm incorporating pruning methodologies

The algorithm of Cl-GBI assimilating the pruning steps is shown in Fig. 6, in which Step 2.1–2.3 correspond to the Step 2 in the algorithm shown in Fig. 1. Note that τ is updated at Step 2.2 if possible.

4   Experimental Evaluation

To evaluate the performance of the proposed pruning strategies, we implemented the algorithm shown in Fig. 6 on PC (CPU: Pentium 4 3.2GHz, Memory: 4GB,

Table 1. Size of graphs of the synthetic dataset and the hepatitis dataset

class	Synthetic		Hepatitis	
	positive	negative	R	N
Number of graphs	150	150	38	56
Average number of nodes in a graph	50	50	104	112
Total number of nodes	7,524	7,502	3,944	6,296
Number of distinct node labels	20		12	
Average number of links in a graph	498	495	108	117
Total number of links	74,631	74,198	4,090	6,577
Number of distinct link labels	20		30	

Table 2. Runtime and maximum information gain found (MIG) for the synthetic dataset

Pruning setting	$L = 5$		$L = 10$		$L = 15$	
	time (s)	MIG	time (s)	MIG	time (s)	MIG
no pruning	1,286	0.2042	2,850	0.2042	3,701	0.2042
only pre-pruning	1,242	0.2042	2,048	0.2042	2,467	0.2042
only post-pruning	597	0.2042	788	0.2042	1,183	0.2042
pre/post-pruning	602	0.2042	675	0.2042	1,007	0.2042

OS: Fedora Core release 3) in C++, and applied it to both synthetic and real-world datasets. The synthetic dataset was generated in a random manner same as [10], and divided into two classes of equal size, "positive" and "negative". Then, we embedded some discriminative subgraphs only in graphs of the class "positive". On the other hand, as a real-world dataset, we used the two classes in the hepatitis dataset, *Response* and *Non-Response*, denoted by R and N, respectively [11]. R consists of patients to whom the interferon therapy was effective, while N consists of those to whom it was not effective. We converted the records of each patient into a directed graph in the same way as [11]. The statistics on the size of graphs in both datasets are shown in Table 1.

Since we adopted the information gain as the criterion to evaluate the discriminativity of subgraphs, C was set to "information gain". Then, we observed the runtime and maximum information gain of enumerated subgraphs under the following four pruning settings: applying no pruning, applying only the pre-pruning, applying only the post-pruning, and applying both of them. The other parameters were set as follows: $b = 5$ and $L = 5, 10, 15$. As for θ, we did not set any requirement to observe only the effect of the proposed pruning methods. The initial value of τ was set to 0.

Table 2 shows the experimental results for the synthetic dataset. It is obvious that applying both pruning methods was succeeded in significantly reducing the runtime. In fact, the reduction ratios in runtime are about 55% for $L = 5$ and about 75% for $L = 10$ and 15. It is noted that the post-pruning is more effective than the pre-pruning. This is because the post-pruning that uses the actual

Table 3. Runtime and maximum information gain found (MIG) for the hepatitis dataset

Pruning setting	$L = 5$		$L = 10$		$L = 15$	
	time (s)	MIG	time (s)	MIG	time (s)	MIG
no pruning	66	0.1890	534	0.1890	1,894	0.1890
only pre-pruning	42	0.1890	380	0.1890	1,430	0.1890
only post-pruning	35	0.1890	140	0.1890	362	0.1890
pre/post-pruning	34	0.1890	137	0.1890	353	0.1890

frequency of a subgraph is more restrictive than the pre-pruning that uses the maximally estimated frequency.

As well, similar tendency is found from the results for the hepatitis dataset shown in Table 3. Indeed, applying both pruning methods reduces the runtime by about 50% for $L = 5$, about 75% for $L = 10$, and about 80% for $L = 15$. These results show that the proposed pruning strategies are useful even under a realistic problem setting, although the effectiveness of the pre-pruning is limited also in this case.

5   Conclusions

In this paper, we proposed two kinds of pruning strategies based on the upper-bound of information gain for Cl-GBI to find discriminative subgraphs from graph structured data. These pruning strategies can exclude unfruitful subgraphs from the search space of Cl-GBI effectively. The experimental results on both synthetic and real world datasets showed their effectiveness. In fact, the application of them reduced the runtime of Cl-GBI by at most 80% on a real world dataset. Although we consider the 2-class problem in this paper, the proposed pruning strategies can be easily extended to the case that there are more than 2 classes. In addition, similar pruning strategies can be defined for the other criteria if they are convex as well as information gain.

For future work, we plan to conduct more sophisticated experimental analyses of the performance of these pruning methods. In addition, we need to seek a method to more tightly estimate the frequency of subgraphs to improve the effectiveness of the pre-pruning. One promising approach might be the use of the information of edges adjoining to each node composing a node pair.

Acknowledgments. This work was partly supported by a grant-in-aid for Young Scientists (B) No. 19700145 from MEXT, Japan.

References

1. Cook, D.J., Holder, L.B.: Substructure discovery using minimum description length and background knowledge. Journal of Artificial Intelligence Research 1, 231–255 (1994)
2. Yoshida, K., Motoda, H.: Clip: Concept learning from inference patterns. Artificial Intelligence 75(1), 63–92 (1995)

3. Matsuda, T., Motoda, H., Yoshida, T., Washio, T.: Mining patterns from structured data by beam-wise graph-based induction. In: Lange, S., Satoh, K., Smith, C.H. (eds.) DS 2002. LNCS, vol. 2534, pp. 422–429. Springer, Heidelberg (2002)
4. Yan, X., Han, J.: gSpan: Graph-based structure pattern mining. In: 2002 IEEE International Conference on Data Mining (ICDM 2002), pp. 721–724 (2002)
5. Inokuchi, A., Washio, T., Motoda, H.: Complete mining of frequent patterns from graphs: Mining graph data. Machine Learning 50(3), 321–354 (2003)
6. Kuramochi, M., Karypis, G.: An efficient algorithm for discovering frequent subgraphs. IEEE Trans. Knowledge and Data Engineering 16(9), 1038–1051 (2004)
7. Nguyen, P.C., Ohara, K., Motoda, H., Washio, T.: Cl-gbi: A novel approach for extracting typical patterns from graph-structured data. In: Ho, T.-B., Cheung, D., Liu, H. (eds.) PAKDD 2005. LNCS (LNAI), vol. 3518, pp. 639–649. Springer, Heidelberg (2005)
8. Quinlan, J.R.: Induction of decision trees. Machine Learning 1, 81–106 (1986)
9. Morishita, S., Sese, j.: Traversing itemset lattices with statical metric pruning. In: 19th ACM SIGACT-SIGMOD-SIGART Symposium on Principles of Database Systems, pp. 226–236 (2000)
10. Nguyen, P.C., Ohara, K., Mogi, A., Motoda, H., Washio, T.: Constructing decision trees for graph-structured data by chunkingless graph-based induction. In: Ng, W.-K., Kitsuregawa, M., Li, J., Chang, K. (eds.) PAKDD 2006. LNCS (LNAI), vol. 3918, pp. 390–399. Springer, Heidelberg (2006)
11. Geamsakul, W., Yoshida, T., Ohara, K., Motoda, H., Yokoi, H., Takabayashi, K.: Constructing a decision tree for graph-structured data and its applications. Fundamenta Informaticae 66(1-2), 131–160 (2005)

A Novel Classification Algorithm Based on Association Rules Mining

Bay Vo[1] and Bac Le[2]

[1] Institute for Education Research, Ho Chi Minh City, Vietnam
[2] Faculty of Computer Sciences – Natural Sciences University
National University of Ho Chi Minh City, Vietnam
vdbay@ier.edu.vn, lhbac@fit.hcmuns.edu.vn

Abstract. The traditional methods for mining classification rules such as heuristics or greedy methods only generate the rules that are too general or overfitting to do with the given database. Thus, they introduce high error ratio. Recently, a new method of mining classification rules is proposed: classification rules mining based on association rules (CARs). It is more advantageous than the traditional methods in that it removes noise and therefore the accuracy is higher. In this paper, we propose ECR-CARM algorithm. It is based on ECR-tree to find all CARs. Besides that, it is necessary for redundant rules pruning and rules reducing to gain the smaller rules set (i.e., reducing the time of identifying the class of new cases and increasing the accuracy). We also develop property to fast prune rules.

Keywords: Association Rules, Classifier, Classification Association Rules, ECR-tree.

1 Introduction

The aim of classification is finding set of rules in database to form accurate classifier. Association rules mining is finding all rules satisfying the given support and confidence threshold. For association rules mining, the target attribute is not pre-determined. However, the target attribute must be pre-determined in classification rules mining.

Recently, there are many methods for finding classification rules based on association rules such as: CPAR [11], CMAR [8], CBA [5], MMAC [10], MCAR [9]. Experiments show that CARs is more accurate than the traditional methods such as C4.5 [8].

At the present, there are two main approaches to find CARs:

1. Classification based-on Apriori algorithm [5]: this approach generates 1-itemset candidates, then calculates the support for finding itemsets satisfying *minSup*. From the frequent 1-itemset, it generates 2-itemset candidates until there is no more candidate generated. This method generates a lot of candidates and scans the database many times so it is time-consuming.
2. Classification based-on FP-tree [8]: this method is proposed by W. Li, J. Han, J. Pei (2001). Its advantage is to scan the database only two times, and uses FP-tree

D. Richards and B.-H. Kang (Eds.): PKAW 2008, LNAI 5465, pp. 61–75, 2009.
© Springer-Verlag Berlin Heidelberg 2009

to compress data and uses tree-projection to find frequent itemsets. This method does not generate candidates. FP-tree structure does not store itemset's class, so it is hard to expand for mining CARs.

In this paper, we present a new method of mining classification rules based on association rules aiming to find out all classification rules existing in the database that satisfies *minSup* and *minConf*.

Our contributions of this paper are as follows:

- We propose ECR-tree (Equivalence Class Rule tree) structure.
- We propose classification association rules algorithm based on ECR-tree.
- We develop algorithms to fast reduce redundant rules.

In section 2, we introduce concepts related to CARs. The main contribution is in section 3. In this section, we develop ECR-tree structure and ECR-CARM algorithm to find CARs. Section 4 represents how to prune redundant rules, and proposes a method for pruning redundant rules faster. We present a method for rules reducing follow in CMAR [8] in section 5. In section 6, we present experiments results. The last section consist of our conclusion and future work.

2 Concepts

2.1 Definitions [8, 10]

Let D be the set of training data with n attributes A_1, A_2, ..., A_n and $|D|$ rows (cases). Let $C = \{c_1, c_2, ..., c_k\}$ be a list of class labels. Specific values of attribute A_i and class C are denoted by lower case a and c, respectively.

Definition 1. An itemset includes set of pairs which are an attribute and a specific value for each attribute in the set, denoted $< (A_{i1}, a_{i1}), (A_{i2}, a_{i2}), ... (A_{im}, a_{im})>$.

Definition 2. A rule r has the form of $<(A_{i1},a_{i1}), ..., (A_{im}, a_{im})> \rightarrow c_j$, where $<(A_{i1},a_{i1}), ..., (A_{im}, a_{im})>$ is an itemset and $c_j \in C$ is a class label.

Definition 3. The actual occurrence $ActOcc(r)$ of a rule r in D is the number of rows of D that match r's condition.

Definition 4. The support of r, denoted $Supp(r)$, is the number of rows of D that match r's condition, and belong to r's class.

Definition 5. The confidence of r, denoted $Conf(r)$, is defined as:

$$Conf(r) = \frac{Supp(r)}{ActOcc(r)}$$

Consider r = $\{<(A, a1)> \rightarrow y\}$ from the database in Table 1, we have:

$$ActOcc(r) = 3$$

$$Supp(r) = 2 \text{ and } Conf(r) = \frac{SuppCount(r)}{ActOcc(r)} = \frac{2}{3}.$$

Table 1. Training dataset

OID	A	B	C	class
1	a1	b1	c1	y
2	a1	b2	c1	n
3	a2	b2	c1	n
4	a3	b3	c1	y
5	a3	b1	c2	n
6	a3	b3	c1	y
7	a1	b3	c2	y
8	a2	b2	c2	n

2.2 Associative Classification

As association rules mining, itemsets satisfying *minSup* are called frequent itemsets. If a frequent itemset has only one item, it is called single item. For example, if *minSup* = 20% we have single items: <(A, a1)>, <(A, a3)>, <(B,b2)>, <(B,b3)>, <(C, c1)>.

3 ECR-CARM Algorithm (Classification Based-on Association Rules Mining)

We propose ECR-CARM, a novel classification algorithm based-on association rules. It computes the intersection among *Obidsets* for computing the support faster, so it only scans the database one time.

3.1 Definitions

a. Definition 6 - Obidset (Object identifier set)
Obidset(X) is a set of object identifications in D that matches X.
Example:

   X1 = <(A,a2)> then *Obidset*(X1) = {3, 8} or 38
   X2 = <(B, b2)> then *Obidset*(X2) = 238
   X3 = <(A,a2), (B, b2)> then *Obidset*(X3) = *Obidset*(X1) ∩ *Obidset*(X2) = 38

b. Definition 7 – Equivalence Class [13]
Let I be a set of items, and $X \subseteq I$. A function $p(X,k) = X[1:k]$ as the k length prefix of X and a prefix-based equivalence relation θ_K on itemsets is defined: $\forall X, Y \subseteq I, X \equiv_{\theta_k} Y \Leftrightarrow p(X,k) = p(Y,k)$. That is, two itemsets are in the same class if they share a common k length prefix.
c. Remark: Supp(X) = |Obidset(X)|
Since *Obidset*(X) is a set of object identifications in D that match X, so *Supp*(X) = |*Obidset*(X)|.

3.2 ECR-Tree (Equivalence Class Rule-Tree)

a. Vertex: Include list of elements. Each element is a triple of
    itemset.
    Obidset.
    Number of *Obidsets* belong to class c_i (i=1..k, where k is a number of target attribute values), denoted $count_i$ $\forall i \in [1,k]$.
The triple is denoted:

$$[\{values\}\ (count(x \in X | decision(x) = c_i) \forall i \in [1,k])]$$
$$\{X = x_{i1}, x_{i2}, \ldots, x_{it}\}$$

In Figure 1, $a1(2,1)$ notation, i.e., the objects 1,2 and 7 contain a1 with two objects
$\{127\}$
have decision "y" and one has decision "n".

b. Arc: Connect between two vertexes such that their attributes are in same equivalence class and have parent-child relationship.

Consider Figure 1, the vertex containing A that connects to AB, AC because A is child of AB, AC and both of them are in same equivalence class with prefix A.

3.3 Algorithm

Input: *minSup, minConf*, and root node *Lr* of ECR-tree contains child vertexes such that their elements contain a frequent 1-itemsets.
Output: CAR contains classification rules of database.
Method:

```
ECR-CARM(Lr, minSup, minConf, m = 0)
   CAR = ∅
   for all lᵢ ∈ Lr do
        ENUMERATE_RULES(lᵢ, minConf )
        Pᵢ = ∅
        for all lⱼ ∈ Lr with j > i do
             JOIN( Pᵢ, lᵢ, lⱼ, minSup, m)
        ECR-CARM(Pᵢ, minSup, minConf, m+1)
   ENUMERATE_RULES(l, minConf )
     for all O ∈ l do
        pos = 1
        for all countᵢ ∈ O.count do
          if countᵢ > O.countₚₒₛ then
              pos = i
        conf = O.countₚₒₛ / | O.Obidset|
        if conf ≥ minConf then
             CAR=CAR∪{O.itemset → cₚₒₛ (|O.Obidset|,conf)}
   JOIN(Pᵢ, lᵢ, lⱼ, minSup, m)
     for all Oᵢ ∈ lᵢ do
        for all Oⱼ ∈ lⱼ do
```

if $p[O_i.itemset, m] = p[O_j.itemset, m]$ then
   $O.itemset = O_i.itemset \cup O_j.itemset$
   $O.Obidset = O_i.Obidset \cap O_j.Obidset$
   if $|O.Obidset| \geq minSup$ then
      Add O to P_i

Algorithm 1. ECR-CARM

The input parameter m (start with zero) indicates that the equivalence classes have the same first frequent m-itemset. ECR-CARM uses this parameter m because the line 3 of the JOIN function checks whether $p[O_i.itemset, m] = p[O_j.itemset, m]$ or not. If it is true, then itemsets of O_i and O_j share frequent m-itemsets, ECR-CARM will generate a new element O by union between itemsets of O_i and O_j, if the support of O.itemset satisfies *minSup*, that is $|O.Obidset| \geq minSup$, then add O into node P_i. Assume l_i is an element that contains frequent k-itemset, then P_i contains all frequent (k+1)-itemsets with the prefix $l_i.itemset$. Consider the function ENUMER-ATE_RULES $(l, minSup, minConf)$: this function generates all rules from elements in node l. Firstly, it scans overall elements in l, then finds out the class label containing maximum number of objects in every element. Secondly, it computes the confidence of rule, if the support and the confidence satisfy *minSup* and *minConf*, then add the new rule into CARs.

For example: consider node A in ECR-tree of Figure 1, we have:

- Element a1(2, 1): i.e., A = a1 includes 3 objects, which have 2 objects belong to class "y", one belongs to class "n" $\Rightarrow$ pos = 1 (count$_1$ > count$_2$). Thus, the support of rule is 2 and the confidence is $\frac{2}{3}$.
- Element a2(0, 2): contains 2 objects and both of them belong to "n". Thus, the support of the generated rule is 2 and its confidence is 100%.

Continue to ECR-CARM, each child of root node will join to the rest nodes behind it to generate a new equivalence class and repeat until there is no any equivalence class generated.

3.4 Illustration

With *minSup* = 20% and *minConf* = 60%. ECR-tree shows classification rules finding process as Figure 1.

Firstly, the root node ({ }) contains the child-nodes that have one attribute include A, B, C (called A, B, C nodes) in which node A contains 3 elements:

- a1 belongs to O-ID that is the 127 with 2 "y" decisions and one "n" decision.
- a2 belongs to O-ID that is the 38 with zero "y" decision and two "n" decisions.
- a3 belongs to O-ID that is the 456 with two "y" decisions and one "n" decision.
- Similar to B and C.
- Consider to generating node AB: each element of node AB generated from an element of node A that joins with an element of node B such that their itemsets

have same prefix of m-itemset. Since the child nodes of the root have m = 0, so the prefix is ∅. Thus, the generating AB is only joining between the elements of A and B together (satisfying *minSup*),   and the results only contain a2b2, a3b3 satisfying *minSup*.

- AC, BC nodes were generated as same as AB node.
- Consider to generate node ABC: ABC node generated from AB and AC nodes. It only has AB = a3b3 and AC = a3c1 which have the same prefix a3. Thus, ABC only has an element a3b3c1 with Obidset = 46.

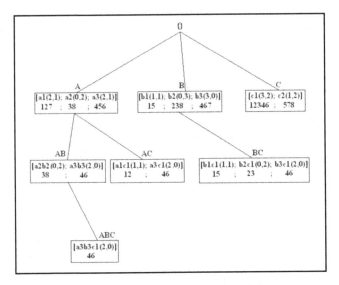

Fig. 1. ECR-tree shows classification rules mining process

Consider to generate rules from the nodes in ECR-tree (follow in recursive order):

- Node A:

  a1: $conf = \frac{2}{3} \geq minConf \Rightarrow$ Rule 1: if A = a1 then class = y $(2, \frac{2}{3})$.

  a2: $conf = 1 \geq minConf \Rightarrow$ Rule 2: if A = a2 then class = n $(2, 1)$.

  a3: $conf = \frac{2}{3} \geq minConf \Rightarrow$ Rule 3: if A = a3 then class = y $(2, \frac{2}{3})$.

- Node AB:

  a2b2: $conf = 1 \geq minConf \Rightarrow$ Rule 4: if A=a2 and B=b2 then class=n $(2, 1)$.

  a3b3: $conf = 1 \geq minConf \Rightarrow$ Rule 5: if A=a3 and B=b3 then class=y $(2, 1)$.

- Node ABC:

  a3b3c1: $conf = 1 \geq minConf \Rightarrow$ Rule 6: if A = a3 and B = b3 and C = c1 then class = y $(2, 1)$.

- Node AC:

  a1c1: $conf = \frac{1}{2} < minConf.$

  a3c1: $conf = 1 \geq minConf \Rightarrow$ Rule 7: if A=a3 and C=c1 then class=y (2, 1).

- Node B:

  b1: $conf = \frac{1}{2} < minConf.$

  b2: $conf = 1 \geq minConf \Rightarrow$ Rule 8: if B = b2 then class = n (3, 1).

  b3: $conf = 1 \geq minConf \Rightarrow$ Rule 9: if B = b3 then class = y (3, 1).

- Node BC:

  b1c1: $conf = \frac{1}{2} < minConf.$

  b2c1: $conf = 1 \geq minConf \Rightarrow$ Rule 10: if B=b2 and C=c1 then class=n (2,1).

  b3c1: $conf = 1 \geq minConf \Rightarrow$ Rule 11: if B=b3 and C=c1 then class=y (2,1).

- Node C:

  c1: $conf = \frac{3}{5} \geq minConf \Rightarrow$ Rule 12: if C = c1 then class = y (3, $\frac{3}{5}$).

  c2: $conf = \frac{2}{3} \geq minConf \Rightarrow$ Rule 13: if C = c2 then class = n (2, $\frac{2}{3}$).

Thus, there are 13 classification rules satisfying $minSup = 20\%$ and $minConf = 60\%$.

4   Redundant Rules Pruning

ECR-CARM generates a lot of rules some of which are redundant rules, because they can be inferred from the other rules. Therefore, it is necessary to prune them. One of the simple methods is represented in [5]. When the candidate k-itemsets are generated, it prunes redundant rules by comparing them with other rules of 1-itemset, 2-itemset, ..., (k-1)-itemset. However, this method is time-consuming because the number of rules are very large. Thus, it is necessary to have more efficient method to prune redundant rules.

4.1   Definitions

a) Definition 8 – Sub-rule

Let two rules R_i and R_j, where R_i: $<(A_{i1}, a_{i1}), ..., (A_{iu}, a_{iu})> \rightarrow c_k$ and R_j: $<(B_{j1}, b_{j1}), ..., (B_{jv}, b_{jv})> \rightarrow c_l$, we say that R_i rule is sub-rule of R_j if it satisfies:

  a. $u \leq v$.
  b. $\forall k \in [1,u]: (A_{ik}, a_{ik}) \in <(B_{j1}, b_{j1}), ..., (B_{jv}, b_{jv})>$.

b) Definition 9 [5]

Given two rules, r_i and r_j, $r_i \succ r_j$ (also called r_i precedes r_j or r_i has a higher precedence than r_j) if

1. The confidence of r_i is greater than that of r_j, or
2. Their confidences are the same, but the support of r_i is greater than that of r_j, or
3. Both the confidences and supports of r_i and r_j are the same, but r_i is generated earlier than r_j

c) Definition 10 – Redundant rules

Let R = $\{R_1, R_2, ..., R_s\}$ be a set of classification rules in database D, R_j is called redundant rule if R contains R_i rule such that: R_i is a sub-rule of R_j and $R_i \succ R_j$ ($i \neq j$).

4.2 Property 1 – Pruning Fast Redundant Rules

If R_i rule has the confidence = 100%, then all rules generated later than R_i and have the same prefix R_i are redundant rules.
Proof:

Let R_S be a set of rules generated later than R_i and have the same prefix R_i. To prove this property, we must prove two conditions:

1. R_i rule is sub-rule of all rules in R_S: obviously.
2. R_i rule has higher precedence than the rules in R_S.
   R_i rule has the confidence = 100%, so the rules in R_S have the confidence = 100% too. Therefore, to prove R_i having higher precedence than the rules in R_S, we must prove that the support of R_i has higher than or equal to the support of each rule in R_S. Therefore, if A $\subset$ B then $Supp(A) \geq Supp(B)$. Since R_i is sub-rule of rules in R_S, the support of R_i is higher than or equal to the support of each rule in R_S.

Thus, R_i rule has higher precedence than the rules in R_S.

4.3 pCARM Algorithm

In this section, we present pCARM, an extension algorithm of ECR-CARM, to prune redundant rules. According to property 1, if the element contains rule with the confidence = 100%, it must be pruned from vertex that contains it. Besides that, if a generated rule with the confidence < 100%, it must be checked whether it is redundant or not.

4.3.1 Algorithm

Input: Root node Lr of ECR-tree contains sub-vertexes such that their elements have frequent 1-itemsets , $minSup$ and $minConf$.
Output: pCARs contains classification rules of database after pruning redundant rules.
Method:

```
pCARM(Lr, minSup, minConf, m = 0)
    pCARs = ∅
    for all lᵢ ∈ Lr do
            ENUMERATE_RULES(lᵢ, minSup, minConf )
            Pᵢ = ∅
            for all lⱼ ∈ Lr with j > i do
                    JOIN( Pᵢ, lᵢ, lⱼ, minSup, m)
```

pCARM(P_i, minSup, minConf, m+1)
ENUMERATE_RULES(l, minSup, minConf)
  for all $O \in l$ do
    pos = 1
    for all $count_i \in O.count$ do
      if $count_i > O.count_{pos}$ then
        pos = i
        conf = $O.count_{pos}$ / $| O.Obidset|$
        r = $\{O.itemset \rightarrow c_{pos} (O.\ count_{pos}, conf)\}$
        if $O.count_{pos} \geq minSup$ then
          if conf = 1.0 then
            pCARs = pCARs $\cup$ r
            delele O from l // Remove O from l
          else
            if conf $\geq minConf$ and NONEREDUND(r, pCAR) then
              pCARs = pCARs $\cup$ r

Algorithm 2. pCARM algorithm

pCARM based on property 1 to prune fast redundant rules, pCARM differs ECR-CARM only in the function ENUMERATE_RULES: The difference is in the case of the confidence = 100%, the function removes the element generated this rule from the node containing it. Thus, this function will not generate any superset that uses the itemset of this rule as prefix. For rules with the confidence < 100%, we need to check the redundancy by comparing to the rules generated earlier according to definition 9.

4.3.2 Illustration

Consider to dataset given in Table 1, where minSup = 20% and minConf = 60%, we have ECR-tree that shows classification rule finding process following by pCARM as figure 2,3 and 4.
The process of rule generating in figure 2 such as:

- Node A:

  a1: $conf = \frac{2}{3} \geq minConf \Rightarrow$ Rule 1: if A = a1 then class = y (2, $\frac{2}{3}$).

  a2: $conf = 1 \geq minConf \Rightarrow$ Rule 2: if A = a2 then class = n (2, 1).

  Because rule 2 has the confidence = 100%, it is necessary to prune the element a2 from node A.

  a3: $conf = \frac{2}{3} \geq minConf \Rightarrow$ Rule 3: if A = a3 then class = y (2, $\frac{2}{3}$).

- Node B:

  b1: $conf = \frac{1}{2} < minConf.$

  b2: $conf = 1 \geq minConf \Rightarrow$ Rule 4: if B = b2 then class = n (3,1).

  Because rule 4 has the confidence = 100%, it is necessary to prune the element b2 from node B.

  b3: $conf = 1 \geq minConf \Rightarrow$ Rule 5: if B = b3 then class = y (3, 1).

Because rule 5 has the confidence = 100%, it is necessary to prune the element b3 from node B.

- Node C:

  c1: $conf = \frac{3}{5} \geq minConf \Rightarrow$ Rule 6: if C = c1 then class = y (3, $\frac{3}{5}$).

  c2: $conf = \frac{2}{3} \geq minConf \Rightarrow$ Rule 7: if C = c2 then class = n (2, $\frac{2}{3}$).

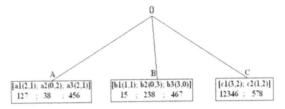

Fig. 2. level 1 in ECR-tree of pCARM

Thus, the ECR-tree after pruning elements:

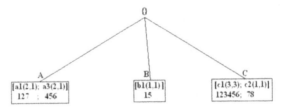

Fig. 3. ECR-tree after pruning elements generating the rules with the confidence = 100%

Next, the elements in A will join elements in B to generate node AB. However, since no any element satisfying *minSup*, so AB is not generated.

In case of joining A and C: there are 2 elements generated. Those are a1c1(1,1) and a3c1(2,1).

Similarity, joining B and C will generate b1c1(1,1).

We have the last ECR-tree:

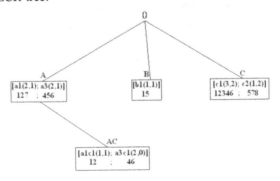

Fig. 4. The last ECR-tree according to pCARM

Node AC generates a rule:

a3c1 has $conf = 1 \geq minConf \Rightarrow$ Rule 8: if A=a3 and C=c1 then class =y (2, 1).

Thus, after pruning redundant rules, we have:
Rule 1: if A = a1 then class = y (2, 2/3).
Rule 2: if A = a2 then class = n (2,1).
Rule 3: if A = a3 then class = y (2, 2/3).
Rule 4: if B = b2 then class = n (3, 1).
Rule 5: if B = b3 then class = y (3, 1).
Rule 6: if C = c1 then class = y (3, 3/5).
Rule 7: if C = c2 then class = n (2, 2/3).
Rule 8: if A=a3 and C=c1 then class =y (2, 1)

5 Building Classifier

Number of rules generated by pCARM method is also very large, so we must have a method to remove some "unnecessary" rules from rules set to help the prediction process faster and more accurate. We base on CMAR [8] to reduce rules. The ideal is: Given a threshold min_δ, the rules covering at least one data row will be gained. The data row will be removed if it is satisfied min_δ rules. The experiment shows that the method is very effective in time, in space, and in accuracy.

5.1 prCARM Algorithm

Input: R contains classification rules of database (CARs or pCARs) and threshold min_δ.

Output: pR is set of rules after pruning by CMAR.

Method:

```
pRCARM()
    SORT_RULES(R)
    count[i] = 0  ∀i∈ [1, N] // N : number of cases
    pR = ∅
    nData = 0
    for i = 1 to |R|  do
        if   nData < N   then
            f = false
            for j = 1 to N do
                if count[j] < min_δ then
                    if LEFT(R[i]) ⊆ data[j]
                        and R[i].class = data[j].class  then
                        count[j] = count[j] +1
                        if count[j] = min_δ then
                            nData = nData + 1
```

$$f = true$$
if f = true then
$$pR = pR \cup R[i]$$

Algorithm 3. Classifier algorithm

At the beginning, it arranges the rules set (R = pCARs or CARs) by decreasing order of precedence, to assure the classifier finds out the high precedence rules. nData is the number of data rows covering by selected rules in pR. In line 12, if the number of rules satisfies the line j (count[j]) = min_δ (line 11), then nData is increased by 1 until nData = N (line 5) . The algorithm is stopped if nData = N or in case of all rules considered (line 4). At line 5, LEFT(R[i]) means that it gets the left side of rule R_i, data[j] is the jth data row in database D. data[j].class in line 6 is the class label of jth case in database D.

5.2 Illustration

Consider the rules that are generated in section 4.3.2 with threshold $min\_\delta = 1$ after arranging by an decreasing order of precedence:

Table 2. The results after arranging

R	Rules		Supp	Conf
R4	If B=b2	then class = n	3	1
R5	if B = b3	then class = y	3	1
R2	if A = a2	then class = n	2	1
R8	if A=a3 and C=c1	then class = y	2	1
R1	if A = a1	then class = y	2	2/3
R3	if A = a3	then class = y	2	2/3
R7	if C = c2	then class = n	2	2/3
R6	if C = c1	then class = y	3	3/5

Rule R4: this rule covers the data rows 2,3,8 so it is kept and the correlative rows are removed.
Rule R5: this rule covers the data rows 4,6,7 so it is kept and the correlative rows are removed.
Rule R2: This rule does not cover any data row (after removed) so it is not kept.
Rule R8: This rule does not cover any data row so it is not kept.
Rule R1: this rule covers the data row 1 so it is kept and the correlative row is removed.
Rule R3: this rule covers the data rows 5 so it is kept and the correlative row is removed.
Finally, all data rows are pruned and there are 4 rules in Table 3

Table 3. The result rules

R	Rules		Supp	conf
R4	If B=b2	then class = n	3	1
R5	if B = b3	then class = y	3	1
R1	if A = a1	then class = y	2	2/3
R3	if A = a3	then class = y	2	2/3

6 Experiments

The demo is coded by C# in Visual Studio.NET 2003, Windows XP operating system. The experimental results are tested in databases getting from UCI Machine Learning Repository. The datasets are run in Pentium III PC, 1GHz, 384MB.

The number of rules generated in dataset is very large. We also tested by datasets with many different *minSup* values in the same *minConf* = 50%, and found that the larger databases in number of attributes, the more rules generated and the time is longer.

Table 4. The experimental datasets [6]

Dataset	#attr	#class	#record
Breast	11	2	699
German	20	2	1000
led7	7	10	3200
Lymph	18	4	148
tic-tac	9	2	958
Vehicle	18	4	846
Zoo	16	7	101

Table 5. Compare the time and the accuracy of classification between CARs, pCARs and pRCARs

database	minSup (%)	CARs		pCARs			pRCARs		
		#rules	time(s)	#rules	time(s)	accur.	#rule	time(s)	accur.
breast	1	5586	0.33	483	0.18	95.8	78	0.24	95.51
	0.5	9647	0.4	834	0.19	95.94	80	0.31	95.94
	0.3	13870	1.01	1728	0.24	95.94	81	0.66	95.8
	0.1	496808	49.52	4554	0.6	95.94	233	3.26	95.8
German	10	3379	1.02	761	1.24	70	126	1.46	71
	5	19343	2.63	3460	8.79	70	205	10.05	71.7
	3	61493	5.46	9765	73.82	70	222	81.62	72
	1	608759	28.16	61458	2634.2	71.88	337	3141.73	73.5
led7	1	459	0.3	403	0.36	71.81	169	0.57	71.91
	0.5	867	0.32	538	0.36	71.75	194	0.62	71.84
	0.3	907	0.34	539	0.37	71.81	194	0.64	71.84
	0.1	996	0.35	578	0.38	71.94	· 197	0.64	71.97

Table 5. (*Continued*)

lymph	10	16149	1.29	6848	12	82.14	40	17.11	81.43
	8	30911	2.02	12510	44.06	82.86	41	64.27	82.86
	6	73060	3.25	21591	127.25	82.14	43	195.61	81.43
	4	222044	7.37	51721	679.84	81.43	43	1103.84	82.12
tic-tac	1	7193	0.52	3380	1.23	97.79	32	2.77	98.84
	0.5	19575	0.71	5896	3.32	98.11	28	7.22	99.16
	0.3	35763	1.4	9322	7.78	98.11	28	17.43	99.26
	0.1	109033	4.54	19032	13.54	98.11	28	58.46	99.26
vehicle	1	369	0.94	463	0.97	57.38	192	0.99	59.17
	0.5	2240	2.1	2770	2.19	59.52	353	3.17	60.1
	0.3	12542	3.53	6302	4.18	59.88	378	8.85	60.32
	0.2	126221	12.79	22011	19.64	59.88	386	79.6	60.32
zoo	10	116813	3.13	3767	7.25	74	4	8.08	73
	5	274689	6.62	8791	47.27	83	7	54.66	88
	3	469571	11.49	14221	92.87	86	9	122.21	93
	1	1034243	52.33	30661	212.64	86	9	361.5	94
avg		134733	7.28	10871	142.74	79.97	128	190.98	81.11

The time: Since ECR-CARM does not prune redundant rules so its time is shortest. pRCARM is longest because it does both pruning redundant rules and building classifier. However, the number of rules generated by pRCARM is quite small so the prediction time of classes for new cases is faster than two remaining methods.

Accuracy: We use ten-fold cross-validation method to compute the accuracy. The accuracy of CARs and pCARs are the same so we only compare pCARs and pRCARs. Result in Table 5 shows that pRCARs is more accurate than pCARs (or CARs) in all testing times. The average accuracy of all databases follow in pCARs is 79.97% and pRCARs is 81.11%.

Large dataset: We also test the algorithms in large dataset of 57600 rows (7 attributes and 10 classes, *minSup* = 1%), the time for mining is as follow (Table 6):

Table 6. The results with large dataset

CARs (s)	pCARs (s)	pRCARs (s)
2.68	2.67	4.96

7 Conclusion and Future Works

The methods of classification based-on association rules mining is more efficient than traditional methods, so the predictability for accuracy is higher. This method is simpler than traditional ones, because it uses association rules mining. ECR-CARM inherits the strong point of top-down method with 2 significant features: i) not generate candidates.

ii) scan dataset one time and base on intersection between *Obidsets* to compute the support faster. Thus, ECR-CARM can be an effective method for classification rule mining.

We can see the number of rules generated by ECR-CARM is quite large, so the prediction of new cases deals with many barriers (in searching time for rules satisfied new cases). Therefore, we reduce the result set: and only retain the general rules covered dataset. Building classifier will make the accuracy become better and decrease prediction time (as in Table 5).

In future work, we will continue to research pruning methods to be more accurate. Besides that, improving this method for decreasing performance time is also interesting.

References

1. Agrawal, R., Imielinski, T., Swami, A.: Mining Association Rules between Sets of Items in Large Databases. In: Proc. ACM SIGMOD Conference on Management of Data, vol. 1993, pp. 207–216 (1993)
2. Agrawal, R., Srikant, R.: Fast Algorithms for Mining Association Rules. In: Proc. of the 20th VLDB Conference, Santiago, Chile, pp. 487–499 (1994)
3. Coenen, F., Leng, P., Zhang, L.: Threshold Tuning for Improved Classification Association Rule Mining, pp. 216–225. Springer, Heidelberg (2005)
4. Han, J., Kamber, M.: Data Mining: Concept and Techniques, 2nd edn., pp. 285–311. Morgan Kaufmann Publishers, San Francisco (2006)
5. Liu, B., Hsu, W., Ma, Y.: Integrating Classification and Association Rule Mining. In: Proc KDD 1998, pp. 80–86 (1998)
6. http://mlearn.ics.uci.edu/MLRepository.html (Download on 2007)
7. Hu, H., Li, J.: Using Association Rules to Make Rule-based Classifiers Robust. In: The 16th Australasian Database Conference, University of Newcastle, Newcastle, Australia, pp. 47–54 (2005)
8. Li, W., Han, J., Pei, J.: CMAR: Accurate and Efficient Classification Based on Multiple Class-Association Rules. In: Proc ICDM 2001, pp. 369–376 (2001)
9. Thabtah, F., Cowling, P., Peng, Y.: MCAR: Multi-class Classification based on Association Rule. In: The 3rd ACS/IEEE International Conference on Computer Systems and Applications, 2005, pp. 33–39 (2005)
10. Thabtah, F., Cowling, P., Peng, Y.: MMAC: A New Multi-class, Multi-label Associative Classification Approach. In: Fourth IEEE International Conference on Data Mining (ICDM 2004), pp. 217–224 (2004)
11. Yin, X., Han, J.: CPAR: Classification Based on Predictive Association Rules. In: Proc. SIAM Int. Conf. on Data Mining (SDM 2003), pp. 331–335 (2003)
12. Zaki, M.J.: Mining Non-Redundant Association Rules. In: Data Mining and Knowledge Discovery, vol. 9, pp. 223–248. Kluwer Academic Publishers, Dordrecht (2004)
13. Zaki, M.J., Hsiao, C.J.: Efficient Algorithms for Mining Closed Itemsets and Their Lattice Structure. IEEE Transactions on Knowledge and Data Engineering, 462–478 (2005)

Multiple Classification Ripple Round Rules:
A Preliminary Study

Ivan Bindoff, Tristan Ling, and Byeong Ho Kang

University of Tasmania, School of Computing
{ibindoff,trling,bhkang}@utas.edu.au

Abstract. This paper details a set of enhancements to the Multiple Classification Ripple Down Rules methodology which enable the expert to create rules based on the existing presence of a conclusion. A detailed description of the method and associated challenges are included as well as the results of a preliminary study which was undertaken with a dataset of pizza topping preferences. These results demonstrate that the method loses none of the appeal or capabilities of MCRDR and show that the enhancements can see practical and useful application even in this simple domain.

1 Introduction

Traditional expert systems of the 1980s commonly featured either backward or forward chaining inferencing approaches [1-4]. A typical key feature of the inferencing process in these approaches was the ability to treat every piece of information in the system as a fact, including the conclusions drawn from any particular rule. In this way, each rule could draw a conclusion and it was then possible to fire further rules, using that conclusion as a newly known attribute. Fundamentally the design of both Ripple Down Rules (RDR) and Multiple Classification Ripple Down Rules (MCRDR) does not support this process. This in turn makes them unsuitable for certain domains and less expert-friendly, as there may arise situations where the expert wishes to define a rule which is logically based on the conclusion of an existing rule or set of rules, but is unable to do so.

The development of Nested Ripple Down Rules (NRDR) by Beydoun and Hoffman partially addressed this situation in the case of single classification problems, and the idea of creating multiple NRDR structures for each possible classification has been suggested to handle multiple classification problems [5]. However to date no fully realized multiple classification, true inferencing RDR solution is known to exist. This is what this paper seeks to address.

2 Past Systems

There is a vast wealth of literature concerning Expert Systems that is well known and understood by this community. However, some small discussion of this literature is required to put this problem in context so a brief discussion of traditional expert systems and a slightly more thorough explanation of RDR are included.

D. Richards and B.-H. Kang (Eds.): PKAW 2008, LNAI 5465, pp. 76–90, 2009.

2.1 Traditional Expert Systems

As was touched upon in the introduction, traditional expert systems of the 1980s and early 90s were generally comprised of a knowledge base, an inference engine and an expert system shell through which to interact with the system. The knowledge base typically consisted of a set of simple IF-THEN rules, which comprised a set of facts to define their condition and produced a single fact as a conclusion. The inference engine was generally either a forward or backward chaining inference engine. A forward chaining inference engine would require a set of all facts and execute the rules accordingly until it reached a conclusion, while a backward chaining inference engine would have a set of goals (conclusions) it wishes to satisfy and work backwards, asking the user for more facts to determine which, if any, of the goals are supported by the evidence [3].

The major flaw that was identified with traditional expert systems was the complications which arose when trying to keep the knowledge base maintained for a reasonable period of time. It was often found that although adding new knowledge was initially a fast and easy process, it became increasingly time consuming as the knowledge base grew, since the knowledge engineers responsible had to manually verify that the new knowledge they attempted to add didn't invalidate the existing knowledge base [2, 4, 6].

2.2 Ripple Down Rules

After suffering similar experiences in the development and maintenance of the GARVAN-ES1 expert system, Compton and Jansen proposed RDR as a new approach for developing expert systems in which rules could be incrementally added as the need arose. Experts would be able to directly provide their *justifications* for the new rule, rather than trying to elaborate some more complex description of how a particular rule for a particular circumstance might be designed [7].

To allow this the RDR knowledge base is structured as a binary tree, with each node comprising of a rule (a set of conditions) and a conclusion. Starting from the root of the tree the inference process attempts to traverse the tree, going down one path if the rule condition was true, and another path if the rule condition was false, until it reaches the end. The last node whose condition is found to be true is considered to have the correct conclusion [8, 9].

To ensure the validity of the knowledge base is maintained every time a rule is added to the system, the case it was added for is stored alongside it as an associated cornerstone case. When the expert seeks to add a new rule the system prompts them to select conditions for their new rule such that the cornerstone case of the rule's parent does not also fire on the new rule, while still giving the current case the correct classification.

Unfortunately RDR was found to have a substantial knowledge acquisition requirement in domains where multiple correct classifications for a single case are common, including the PEIRS (Pathology Expert Interpretive Reporting System) on which RDR was first practically assessed [10]. The issue that arises in these situations is that in order to represent multiple classifications the expert must define a "compound classification", which is a new classification which essentially combines two or more other classifications into one. The repercussion of this is that the expert is then required to define far more rules in order to encapsulate all the plausible combinations [10-13].

2.3 Multiple Classification Ripple Down Rules

To address the problems associated with compound classifications MCRDR was developed by Kang and Compton. As an approach it sought to maintain the philosophy and general strategies that made RDR valuable, but augment it with the ability to produce multiple classifications in a sensible fashion.

To achieve this goal MCRDR alters the underlying binary tree structure and inferencing process of RDR, but preserves the exception-based, context sensitive incremental knowledge acquisition approach of RDR, as well as its innate knowledge base verification and validation [12, 13].

MCRDR uses an n-tree structure, with each node consisting of a set of conditions, a classification, and references to all its children nodes, if any. A root node is also present which has no conditions and can be considered to always be true.

If the expert using the system encounters a case where there is a missing classification, they are required to select the missing classification and a set of conditions which lead to the classification. This new classification will then be added to the root node of the tree.

If the expert encounters a case which has a classification fire incorrectly, the expert selects this classification and is asked to provide the correct classification (which may be simply a NULL classification, to simply stop the existing classification for firing, or another classification altogether) and their justification for why this classification is wrong (selecting conditions for the exception rule). The system will then add this new node as a child to each node which provided the incorrect conclusion on the case.

Through these two processes it is possible for the expert to gradually build and refine the knowledge base, simply by progressively examining cases and attempting to ensure the system classifies them correctly.

With single classification RDR every rule has an associated cornerstone case. With MCRDR it is possible to have many cornerstone cases that apply for any given new rule. Essentially any case which has previously been used to create a rule and which would fire with the new rule that the expert is trying to define can be considered to be a cornerstone case for the current new rule. For the system to allow the expert to successfully create a new rule the expert must either ensure that their new rule will fire on no cornerstone cases (by selecting appropriate "differences" for conditions) or they must concede that the cornerstone cases which do fire *should* have fired on the new rule, and the expert simply missed this fact previously.

2.4 Nested Ripple Down Rules & Repeat Inference MCRDR

NRDR extends RDR to make it suitable for recursive style planning and search domains, such as those encountered in classic expert systems such as XCON, and classic AI problems such as Chess playing. To do this, as the name implies, it nests RDR trees such that the result of one tree can lead into another RDR tree in which that result is known to be true. In this way it can be understood that NRDR is capable of *intermediate* conclusions. This approach has demonstrated some success in suitable domains [14, 15].

However, NRDR like RDR is a single classification method. It has been suggested that to produce multiple classifications one might simply employ multiple NRDR trees, where each tree represents the reasoning for one particular classification that the expert desires.

It is understood that a similar approach to this has been tested (sometimes) under the name Repeat Inference Multiple Classification Ripple Down Rules (RIMCRDR), which is essentially the NRDR strategy but with a differing approach to handling the potential cycles that can occur with this type of inferencing. NRDR resolves conflicts by explicitly detecting them and asking the expert to do secondary refinements while RIMCRDR does not allow conflicts to arise as rules are processed in strict chronological order and assertions can not be undone [14-17].

This approach is seen by this author as suitable for some problems, but lacking some of the finesse and extensibility that an expert might desire in more complex domains. For example, using this approach it is difficult to see how the expert might (without some considerable effort on their behalf, which is reminiscent of the compound classifications problem of using RDR for multi-class problems) define a rule which is based on two or more conclusions being present on the case at the same time. Also, by introducing strict chronological order RIMCRDR would appear to be making some throwbacks to the knowledge base maintenance problems of the past. Further to this, by stating that assertions can not be undone you are removing the possibility of making an exception to a node that is dependent on an intermediate conclusion to fire, which can be seen to be a large restriction in itself (for example, a stopping or replacement rule that only fires after another classification is added to the result set, and retroactively stops/replaces another classification which has already been added to the result set previously).

3 Methodology

In an attempt to overcome the limitations of RIMCRDR a recursive inferencing approach to MCRDR was considered, rather than a repeat inferencing approach to NRDR. Once the details of this approach were determined it was then given an initial trial on a pizza preferences dataset that was acquired through the use of an online survey.

3.1 Multiple Classification Ripple Round Rules

To enable the extension of MCRDR to allow rules based on the presence of other classifications it was deemed necessary to revisit the design of the underlying tree structure. A proposal of this approach was published previously, but has since been considerably fleshed out after experience in practice [18], and dubbed Multiple Classification Ripple Round Rules (MCRRR) by the author.

The resultant structure is no longer a tree in concept, but more a graph. However, the effects of the tree structure are preserved entirely, since the references back to other nodes are stored and handled differently to the references which denote the child/parent relationship.

The key addition to enable this effect is the addition to each node of a set of zero or more switches, which can denote any classification the system can produce. These switches are turned on (incremented by 1) whenever the classification they represent is added to the result list, and, importantly, turned off (decremented by 1) whenever a classification is removed from the result list. Whenever a switch is changed the node it applies to is reconsidered. If it was turned off (decremented back to 0) and the node has *already* been added to the result list, the classification is then removed. If it was turned on (incremented) and *all* the switches that apply to that node are now activated (>0) then the node is considered as normal in MCRDR, firing if its conditions are also satisfied, or firing a child node if it is satisfied as well. The resultant structure can be seen visually in Figure 1. Please note that the rule numbers provided in the diagram are for illustrative purposes only and that the order of rule addition and inferencing has no impact on the result with this method.

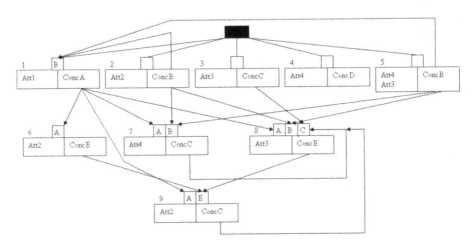

Fig. 1. The graph-like Multiple Classification Ripple Round Rules structure, where the left larger left box represents the normal conditions of the rule, the smaller boxes above represent any conclusion switches which must be turned on for the rule to fire, and the larger right box represents the conclusion of the current rule.

Using the Knowledge Base represented by Figure 1 we can consider a sample case that has Att1, Att2, and Att3 present.

> Rule 1 will be considered first, but will fail to fire as switch B (which represents ConcB) is not yet activated.

> Rule 2 will then be considered, it has no switches and Att2 is satisfied so ConcB will be added. This will result in the B switches for Rule1, Rule 7 and Rule 8 being turned on.

> > Rule 1 now has all its switches active, so Att1 will be considered and ConcA will be added which will in turn, turn on the A switch for Rules 6, 7, 8 and 9.

> > Rule 7 is considered (as a result of the B switch being turned on), and since both its switches are now turned on it will be evaluated, but will fail since Att4 is not true.

> > Rule 8 will be considered, but will not fire as its C switch is off.

> ➤ Rule 6 will be considered and will fire since its switch is on, and Att2 is true. Adding ConcE to the result set will trigger the E switch on Rule 9 to turn on.

> ➤ Rule 7 will be considered again (as a result of the A switch being turned on) but is marked as having already fired so will not be evaluated.

> ➤ Rule 8 is considered, but fails since the C switch is still off.

> ➤ Rule 9 is considered and fires since both the A and E switches are active and Att2 is true. ConcC is added which turns on the C switch on Rule 8.

>> ➤ Rule 9 is considered, but has already fired so will not be re-evaluated.

>> ➤ Rule 8 is considered and fires as the A, B, and C switches are on and Att3 is true. ConcE is added, so each E switch will now be incremented to 2, but will not result in any new nodes becoming active.

> ➤ Rule 3 is considered, and fires adding ConcC to the result set and incrementting each C switch to 2, but not activating any new nodes.

> ➤ Rule 4 is considered, but fails since Att4 is not true.

> ➤ Rule 5 is considered, and also fails due to Att4.

The outcome of this inferencing process is that ConcA, ConcB, ConcC and ConcE are provided back to the user, with the result list containing ConcB, ConcA, ConcE, ConcC, ConcE, ConcC in that order. The fact that any given classification can be reached through multiple paths in MCRDR is important here as can be seen in the above example, where ConcC and ConcE both have a "second chance", whereby their associated switch can be decremented in the event of a conclusion being retroactively removed but their switches can still remain on.

The other thing the reader may notice in the above example is that nodes can be re-visited multiple times without need, particularly in instances where the node has multiple switches. However, since it takes virtually no computational time to assess whether the node's set of switches are all on or all off, and the rest of the node evaluation process is short-circuited if this process fails, this is of no real concern.

To reduce complexity the above example does not include any exceptions and does not illustrate the potential for infinite regress (cycles) that this structure exhibits. However, these situations have been considered and will be discussed here.

3.1.1 Exceptions and Cycles

By allowing exceptions which can use existing conclusions as conditions for firing it is entirely possible to introduce a cycle situation such as the one shown in Figure 2, as it now becomes possible for a conclusion to be added and then retroactively removed by an exception which becomes true at a later stage in the inferencing process. The strategy that has been employed to prevent this undesirable circumstance is in some ways similar to that employed by NRDR, and quite different to that employed in RIMCRDR, in that the expert is simply not allowed to create a rule which has the potential to introduce a cycle in the knowledge base. However, this is done in a somewhat intelligent way in that the system does not detect the cycle in progress

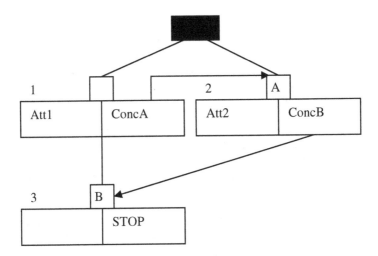

Fig. 2. A simple example of a cycle. Assuming rules are added in order, when the user adds rule 3 they are introducing the potential for an infinite regress with ConcA initially being added, ConcB then being added, as a result of which ConcA is *removed* which in turn removes ConcB which in turn re-adds ConcA and so forth.

and intervene or naively prevent the expert from creating any kind of rule which might potentially cause a cycle. It is instead able to quickly assess any new rule that is being added and determine whether it is a potential cycle creator.

It does this by adding a further concept to the conclusion structure, being that of a *dependency chain*. This is unlike standard cycle detection in a normal directed graph, as it optimizes the process on two fronts by only looking at the new rule under consideration, and by only considering nodes which have the potential to cause cycles, rather than naïvely searching the entire structure for cycles in the graph (many of which may well have rule combinations that make them impossible to fire at any rate). This chain is a simple list of all conclusions, and the context (true or false) that this conclusion is dependent on in order to fire. So in the example shown in Figure 2, when adding :-

➤ Rule 1 ConcA is dependent on nothing.
➤ When adding Rule 2 ConcB is dependent on ConcA (by virtue of A being a switch).
➤ When adding Rule 3 ConcA becomes dependent on ConcB *not* being present and the STOP conclusion would become dependent on B.

It is at this point that the system can detect a cycle, as it can see a dependency path from ConcB -> ConcA -> !ConcB, where the ! denotes a *not* condition. Hence this process is a simple case of searching down the dependency chains for each conclusion from the point where a dependency chain was just added, in an attempt to add the rule. If you ever encounter an already discovered conclusion in the opposite context to the one it was originally found in, a potential cycle can occur and this rule should not be added. At this point it is a simple matter of alerting the expert that their rule could not be added due to the potential of a cycle, and request that they refine their rule further or find an alternative way to add it. In practice it is not expected that this will occur with much frequency, although with very large knowledge bases with a low number of classifications and poorly understood domains ("bad" experts), it may become a prohibitive annoyance and methods may need to be developed to reduce this problem further.

3.1.2 Cornerstone Cases

Another part of MCRDR which required some modifications of interest is that of how to find cornerstone cases.

The concept of a cornerstone case in MCRDR is principally quite simple, "any case which has been previously used to create a rule and would fire on this new rule" [12, 19]. However, MCRDR has been privy to some optimization on the topic of *finding* cornerstone cases as it is a relatively simple matter to store each cornerstone case that applies to a given node (including the root) and simply recall this whenever adding an exception to that and test each of these to see if they would also fire with the new rule. Unfortunately this approach does not hold true after the modifications necessary for MCRRR, and the author has to date thought of no other way to optimize the process, so a relatively naïve approach is used instead.

This approach involves building a mock of the knowledge base as it would be if you assumed the new rule were already added, then inferring every case which has previously been used to make a rule using this mock knowledge base. Every case which fires on the new rule under consideration is a cornerstone case for the new rule, and must be checked for acceptance or differences by the expert as usual.

This method for determining cornerstone cases does seem much less efficient than the alternative, but when it is considered exactly how efficient the actual inferencing process is, even with very large knowledge bases, then it is realized that having to infer many hundreds of past cases will take no appreciable amount of time. Further to this, it is known that in some past prototype MCRDR systems this approach has been taken by (probably unwitting) students and shown no particular ill effects. Therefore it is not anticipated that this will cause any problem at this stage, although in situations with very large knowledge bases and/or very large case histories it may become prohibitive and require some further attempts at optimization.

4   Results and Discussion

As the expert uses the system every action in the system is carefully monitored and recorded into a log file. Analysis of these log files, coupled with examination of the knowledge base itself are used to produce the results detailed below.

4.1   Test Domain

To provide preliminary testing for this new method a small scale survey was undertaken. A comprehensive list of the ingredients available as pizza toppings at 5 pizza restaurants in the Greater Hobart area was compiled and made available as a web-survey on the Internet, where each participant was asked to select Dislike, Neutral or Like for each ingredient. Some 1640 responses were recorded in the two days the survey was available to the public.

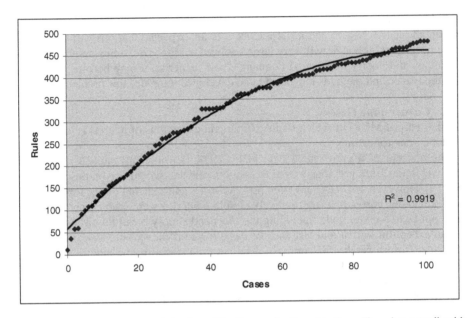

Fig. 3. The growth of the knowledge base. The line can be thought of as a "learning curve", with very heavy learning (steep slope) early on and a much reduced rate of learning at later stages.

The author then compiled the menus of each pizza restaurant which was considered, and by referring to this compilation was able to make suggestions for pizzas from these 5 restaurants to each survey respondent (case). A pizza was considered suitable for the respondent because it included no ingredients they disliked but did include multiple ingredients they did like.

This domain was deemed suitable for testing as it was sufficiently simple for the author (or probably anyone at all) to act as an expert, it was seen to have a large range of classifications (every available pizza from 5 different restaurants), and because it was seen as likely to include situations where it would be desirable to make rules based on conclusions (such as instances where different restaurants had pizzas with much the same list of ingredients as each other).

It is freely conceded by the author that the relative simplicity of this test domain does raise some questions as to its efficacy in thoroughly testing the method. However, multi-class datasets and the associated expert time involved in classifying them are somewhat of a rarity and it was deemed too ambitious to attempt to involve real experts in the project for the first study where the stability, usability and work-flow of the system was still questionable. It was more sought to address those issues through this study than provide truly meaningful insight into the success of such a method. In saying this, there were still worthwhile results to discuss.

4.2 Growth of the Knowledge Base

It was seen that the knowledge base grew in a fashion consistent with MCRDR and RDR systems of the past, with a clear tendency towards flattening off the number of rules per case needed as the system increasingly provided most or all of the conclusions correctly without further refinement. This suggests that there is no particular knowledge acquisition overhead associated with the new method. However, the learning curve is not quite flat suggesting there is still some learning to be done by this system, although it was considered sufficiently trained to terminate the experiment.

The sheer number of rules required (477, far higher than known previous MCRDR systems after a similar number of cases [13, 19]) for only 101 cases analysed suggests the domain was not as simple as expected although this is likely at least partly due to the large number of conclusions that had to be accommodated, as the knowledge base contained 68 conclusions used an average of 3.59 times each. Further supporting the claims that this domain was not as simple as expected is the number of cases seen before a flattening pattern was observed. The number of cases analysed before a flattening in growth can be observed can be seen as a loose indicator of the complexity of the domain, with more complex domains taking longer to flatten. This domain is showing a similar pattern to that which a previous MCRDR system for Medication Review which was considered to be extremely complex showed at around 150 cases [19].

4.3 Accuracy of the Knowledge Base

The accuracy of the knowledge base was determined after each case by calculating the number of conclusions the system correctly identified against the total number of conclusions the case should have had provided (as verified by the expert accepting the case was correctly classified at this stage), in the manner described in Equation 1. It performed as admirably as one would expect a normal MCRDR system to perform, with a rapidly increasing accuracy, although even towards the end of the study there were still some outliers, such as case 87 which was the first lactose intolerant respondent analysed. Their inability to eat cheese threw the system at this stage. Overall however, the system managed to reach into the 90% accuracy range within the 101 case period of the study.

$$\frac{C_f - C_{removed} - C_{replaced}}{C_f + R_a} \tag{1}$$

Where C_f is Conclusions Found, R_a is Rules Added, $C_{removed}$ is Conclusions Removed, $C_{replaced}$ is Conclusions Replaced.

4.4 Time to Create Rules

The time taken to create each rule was quite quick, averaging only 33 seconds where other, perhaps more complicated RDR and MCRDR systems in the past have reported times around the 3 minute mark [9, 13, 19]. Outliers above 100 seconds were inevitably situations where the expert was distracted while creating the rule. Importantly there was no upwards trend in time taken to create rules, which demonstrates that the potential extra overhead in the area of cornerstone cases is not being prohibitive at least up to the 101 case mark, although as the trend-line is visibly completely flat it

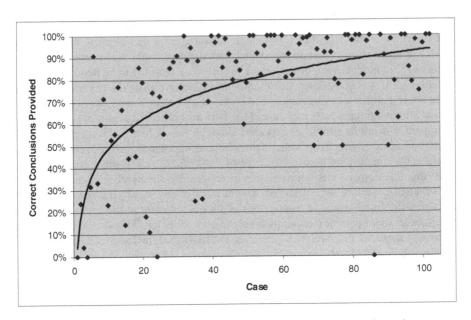

Fig. 4. The percentage of correct conclusions provided by the system for each case

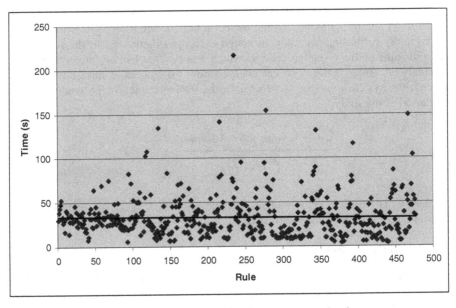

Fig. 5. The time in seconds taken to create each rule

seems unlikely to change substantially, even were considerably more cases to be analysed. This is important as it was known that in order to keep the verification and validity of MCRDR intact in this method a less efficient algorithm for finding cornerstone cases would have to be used. This result shows that this restriction is not prohibitive in terms of wasted expert time, and matches expectations.

4.5 Rules Using Conclusions as Conditions

The expert found 8 situations where it was desirable to base a rule on another conclusion. These 8 rules were subject to some scrutiny, as for each rule it was recorded how many times it was accepted as being correct for a case and how many times it was rejected as wrong (an exception added). The rules were collectively found to be correctly fired 136 times and were only considered wrong once. The relatively low number of rejections by the expert does suggest that they were being conservative when creating these types of rules and only using them when it was quite obvious that the pizza in question was essentially the same as the existing one (all the same ingredients within 1) and this is reflected when looking at the actual rule conditions in question. This result is in many ways disappointing as it was hoped there would be more opportunities to employ this new type of rule in this domain: it is thought that there likely were, but they were overlooked as it was not immediately obvious that a particular situation called for this type of rule.

5 Conclusions

The MCRRR method which is described and tested in this paper had to undergo a range of modifications from classic MCRDR in order to achieve its goal of allowing rules based on other conclusions in a flexible and intuitive manner. Some of these modifications were of slight concern in terms of method efficiency and user friendliness, in particular the less optimal approach to cornerstone case finding and the potential for the extra rules regarding potential cycles becoming prohibitive. However, during the course of this study the first of these concerns did not raise any measurable issues and was not noticed by the expert at all. Although the study is admittedly relatively simple, the simple fact that there was absolutely no indication of inefficiency in the rule creation process to date with 477 rules created suggests that at least standard sized knowledge bases/domains will not be adversely affected by these changes. Unfortunately, it must be conceded that with only 8 rules based on conclusions in the system it is currently impossible to draw any conclusion as to whether the approach used to prevent the introduction of cycles in the system will be prohibitive in practice. However, it should be noted that whether this approach ends up being more prohibitive than necessary or not, it is always guaranteed to be less prohibitive than MCRDR, which does not allow these types of rules at all, since no feature of MCRDR is sacrificed in implementing these extensions. So the only tangible down-side to including this functionality is the (admittedly somewhat substantial) increased complexity of implementation.

Fortunately another, less measurable, set of goals was achieved through the course of this study. It was desired to improve expert workflow, test for correctness and repair any outstanding interface issues. This set of goals does appear to have been satisfactorily achieved, albeit after two experiment re-starts to fix outstanding errors. The author is now satisfied that the system is working as intended, and that the interface can be productively used by an expert to facilitate case classification.

In drawing these few conclusions however, it is far more important at this preliminary stage to be looking toward the future, as much more testing and indeed a more comprehensive list of features are desired.

6 Further Work

It is freely acknowledged by this author that the pizza recommendations experiment outlined in this paper is, although extremely valuable, too simple a domain to make any major claims as to the value of this method. The reasons for this simplicity are outlined above and no further apologies will be made for this, but it is clear that a more comprehensive study is desired. In particular it would be extremely valuable to demonstrate the application of this method to a domain in which MCRDR would be expected to perform poorly and which this method would demonstrate a substantial improvement and a highly tangible value. The intention is not to set MCRDR up for failure and claim some hollow victory, but rather to demonstrate that this method does in fact allow experts to tackle problems that they otherwise would have great difficulty approaching with this kind of method, while still preserving the advantages MCRDR offers. Examples of this type of domain would include:-

- Iterative planning problems, where the decision to place item X in position A impacts on your decisions of where to place item Y.
- Situations where substantial levels of grouping are desired yet are as yet undefined or otherwise difficult/undesirable to hard-code into the system.
  - This can include *many* complex domains where a knowledgeable programmer is not available to hard-code the grouping levels into the system, or domains where the grouping levels are not standardized or yet defined.
- Situations where substantial overlap of rule conditions are likely across different rules.

It should be pointed out that the pizza preferences domain was seen to be one matching that third point, with substantial overlap of rule conditions being likely and it turned out to be quite difficult to manually identify the situations where this overlap would occur. This brings us to another suggestion for further work, whereby the system can automatically recommend situations where using a conclusion as a rule may be valuable. Unfortunately the only obvious way to do this is to wait until the expert has already determined all the rule conditions and then inform them that those conditions were used previously for another rule. If the expert has already gone to the effort of finding those rule conditions, there is little reason to then ask them to change the rule they have defined.

Another feature which would be desirable and is not yet possible with this method is allowing the expert to create rules based on the *absence* of a conclusion. An approach for achieving this has been drafted by the author and it is not believed any further substantial changes to the data structure are required, but it remains to be implemented and tested.

Further to this, it was observed while building the rules for this system that it was extremely common to create a range of exceptions to a rule all with the same goal. Rather than creating multiple exceptions, it was considered that you might create an exception with a range of OR conditions. Or conditions have not previously been available in MCRDR, although it is understood that RIMCRDR did include them also [14]. The historical reasons for this omission appear invalid when considering this method, and it is seen that allowing these conditions might reduce the number of knowledge acquisition steps (and thus rules) that are necessary to achieve the same level of accuracy in the system.

7 Final Words

It is hoped that with the inclusion of some or all of the further work described above this new method to extend the MCRDR approach to a broader range of problems and allow the expert more freedom of knowledge expression will offer a substantial contribution to the RDR community. Initial results are promising, and hopefully are a marker for more to come.

References

1. Miller, R., Pople, H., Myers, J.: Internist-1, an experimental computer-based diagnostic consultant for general internal medicine. The New England Journal of Medicine 307, 468–476 (1982)
2. Buchanan, B., Barstow, D., Bechtal, R., Bennett, J., Clancey, W., Kulikowski, C., Mitchell, T., Waterman, D.: Constructing an Expert System. In: Hayes-Roth, F., Waterman, D., Lenat, D. (eds.) Building Expert Systems, 1st edn., pp. 127–167. Addison-Wesley, Reading (1983)
3. Buchanan, B., Shortliffe, E.: Rule-Based Expert Systems, 1st edn., vol. 1. Addison-Wesley, Reading (1984)
4. Barker, V.E., O'Connor, D.E., Bachant, J., Soloway, E.: Expert systems for configuration at Digital: XCON and beyond. Commun. ACM. 32, 298–318 (1989)
5. Suryanto, H.: Learning and Discovery in Incremental Knowledge Acquisition, Department of Artificial Intelligence, University of New South Wales, Sydney (2005)
6. Bachant, J., McDermott, J.: R1 Revisited: four years in the trenches. Readings from the AI magazine. American Association for Artificial Intelligence, pp. 177–188 (1988)
7. Compton, P., Jansen, R.: A philosophical basis for knowledge acquisition. In: European Knowledge Acquisition for Knowledge-Based Systems, Conference (1989)
8. Compton, P., Jansen, R.: Cognitive aspects of knowledge acquisition. In: AAAI Spring Consortium, Conference (1992)
9. Preston, P., Edwards, G., Compton, P.: A 2000 Rule Expert System Without a Knowledge Engineer. In: AIII-Sponsored Banff Knowledge Acquisition for Knowledge-Based Systems, Conference (1994)
10. Edwards, G., Compton, P., Malor, R., Srinivasan, A., Lazarus, L.: Peirs: A pathologist-maintained expert system for the interpretation of chemical pathology reports. Pathology 25, 27–34 (1993)
11. Compton, P., Kang, B., Preston, P., Mulholland, M.: Knowledge Acquisition without Analysis. In: Knowledge Acquisition for Knowledge-Based Systems, Conference (1993)
12. Kang, B., Compton, P., Preston, P.: Multiple Classification Ripple Down Rules (1994)

13. Kang, B., Compton, P., Preston, P.: Multiple Classification Ripple Down Rules: Evaluation and Possibilities. In: AIII-Sponsored Banff Knowledge Acquisition for Knowledge-Based Systems, Conference (1995)
14. Suryanto, H.: Learning and Discovery in Incremental Knowledge Acquisition, in School of Computer Science and Engineering, University of New South Wales, Sydney (2005)
15. Beydoun, G., Hoffmann, A.: Programming Languages and their Definition. LNCS, pp. 177–186. Springer, Heidelberg (1997)
16. Beydoun, G., Hoffmann, A.: Programming 1980. LNCS, pp. 83–95. Springer, Heidelberg (1998)
17. Compton, P., Cao, T., Kerr, J.: Generalising Incremental Knowledge Acquisition (2004)
18. Bindoff, I., Kang, B.H., Ling, T., Tenni, P., Peterson, G.: Applying MCRDR to a Multidisciplinary Domain. In: Orgun, M.A., Thornton, J. (eds.) AI 2007. LNCS, vol. 4830, pp. 519–528. Springer, Heidelberg (2007)
19. Bindoff, I., Tenni, P., Kang, B., Peterson, G.: Intelligent Decision Support for Medication Review. In: Hoffmann, A., Kang, B.-h., Richards, D., Tsumoto, S. (eds.) PKAW 2006. LNCS, vol. 4303, pp. 120–131. Springer, Heidelberg (2006)

Generalising Symbolic Knowledge in Online Classification and Prediction

Richard Dazeley and Byeong-Ho Kang

School of Information Technology and Mathematical Sciences,
University of Ballarat, Ballarat, Victoria 7353, Australia
School of Computing and Information Systems,
University of Tasmania, Hobart, Tasmania, 7001
r.dazeley@ballarat.edu.au, bhkang@utas.edu.au

Abstract. Increasingly, researchers and developers of knowledge based systems (KBS) have been incorporating the notion of context. For instance, Repertory Grids, Formal Concept Analysis (FCA) and Ripple-Down Rules (RDR) all integrate either implicit or explicit contextual information. However, these methodologies treat context as a static entity, neglecting many connectionists' work in learning hidden and dynamic contexts, which aid their ability to generalize. This paper presents a method that models hidden context within a symbolic domain in order to achieve a level of generalisation. The method developed builds on the already established Multiple Classification Ripple-Down Rules (MCRDR) approach and is referred to as Rated MCRDR (RM). RM retains a symbolic core, while using a connection based approach to learn a deeper understanding of the captured knowledge. This method is applied to a number of classification and prediction environments and results indicate that the method can learn the information that experts have difficulty providing.

Keywords: Hidden context, knowledge based systems, knowledge representation, ripple-down rules, situation cognition.

1 Introduction

Traditionally, knowledge based approaches have been based on the physical symbol hypothesis [1] which is built around the idea that knowledge is a substance that exists. However, after numerous failed systems some researchers have revised these concepts of knowledge and moved towards a situation-cognition (SC) based view. The SC view revolves around the premise that knowledge is generated at the time of its use. This implies that the existence of knowledge is based on the context of a given situation [2, 3]. A few methodologies, such as Formal Concept Analysis (FCA) [4], Repertory Grids [5] and Ripple-Down Rules (RDR) [6], have adopted a weak SC position by including contextual information. These approaches either incorporated the context directly in the knowledge itself or in the structure the knowledge was represented. These methods have been reasonably successful, however, they assume that the context is *a priori*, and therefore, deductive [7]. This assumption leads to static representations of contextual based

D. Richards and B.-H. Kang (Eds.): PKAW 2008, LNAI 5465, pp. 91–108, 2009.
© Springer-Verlag Berlin Heidelberg 2009

knowledge. However, context in certain situations could be considered *a posteriori*, and therefore, inductive [7].

The aim of this paper is to present an algorithm that moves away from these contextually static representations and instead heads towards an intermediate SC [8] view by handling hidden and dynamic contexts. This involves incorporation similar behaviour to the traditional strengths of connection based approaches while still being able to acquire and retain knowledge quickly. The result is a system that learns quickly and is still able to generalise effectively. The results in this paper investigates the method's ability to classify cases quickly in an online environment and to predict continuous values. This notion of a symbolic based system capable of finding hidden contextual information through the generalisation of captured knowledge, led to the notion of combining a Knowledge Based System (KBS) with an Artificial Neural Network (ANN). The KBS selected for use in this paper was MCRDR, as this is currently one of the methodologies most capable of modelling multiple contexts [9].

This paper is broken into three main sections. The first section will provide a background on MCRDR. This is followed by a discussion of the algorithm developed. Lastly, extensive results will be given, detailing the systems ability to discover more knowledge than that provided by the expert in both an online environment and in predicting continuous values.

2 Multiple Classification Ripple-Down Rules (MCRDR)

Ripple-Down Rules is a maintenance centred methodology for a KBS based approach using the concept of fault patching [10] and was first proposed by Compton and Jansen in 1988 [6]. It utilises a binary tree as a simple exception structure aimed at partially capturing the context that knowledge is obtained from an expert. It was assumed that the context was the sequence of rules that had evaluated to provide the given conclusion [6, 11-15]. Therefore, if the expert disagrees with a conclusion made by the system they can change it by adding a new rule. However, the new rule will only fire if the same path of rules is evaluated [13].

Ripple-Down Rules has been shown to be a highly effective tool for knowledge acquisition (KA) and knowledge maintenance (KM). However, it lacks the ability to handle tasks with multiple possible conclusions. Multiple Classification Ripple-Down Rules (MCRDR) aim was to redevelop the RDR methodology to provide a general approach to building and maintaining a Knowledge Base (KB) for multiple classification domains, while maintaining all the advantages from RDR. Such a system would be able to add fully validated knowledge in a simple incremental contextually dependant manner without the need of a knowledge engineer [16, 17].

The new methodology developed by [16] is based on the proposed solution by [12, 13]. The primary shift was to switch from the binary tree to an *n-ary* tree representation. The context is still captured within the structure of the KB and explanation can still be derived from the path followed to the concluding node. The main difference between the systems is that RDR has both an *exception* (true) branch and an *if-not* (false) branch, whereas MCRDR only has exception branches. The false branch instead simply cancels a path of evaluation. Like with RDR, MCRDR nodes each contain a rule and a conclusion if the rule is satisfied. Each, however, can have any number of child branches.

Inference occurs by first evaluating the root and then moving down level by level. This continues until either a leaf node is reached or until none of the child rules evaluate to true. Each node tests the given case against its rule. If false it simply returns, X (no classification). However, if this node's rule evaluates to true then it will pass the case to all the child nodes. Each child, if true, will then return a list of classifications. Each list of classifications is collated with those sent back from the other children and returned. However, if none of the children evaluate to true, and thus they all return X, then this node will instead return its classification. Like with RDR the root node's rule always evaluates to true, ensuring that if no other classification is found then a default classification will be returned.

Knowledge is acquired by inserting new rules into the MCRDR tree when a misclassification has occurred. The new rule must allow for the incorrectly classified case, identified by the expert, to be distinguished from the existing stored cases that could reach the new rule [18]. This is accomplished by the user identifying key differences between the current case and each of the rules' cornerstone cases. A cornerstone case is a case that was used to create a rule and was also classified in the parent's node, or one of its child branches, of the new node being created. This is continued for all stored cornerstone cases, until there is a composite rule created that uniquely identifies the current case from all of the previous cases that could reach the new rule. The idea here is that the user will select differences that are representative of the new class they are trying to create [18].

3 Methodology

The approach developed in this paper is a hybrid methodology, referred to as Rated MCRDR (RM), combining MCRDR with a function fitting technique, namely an artificial neural network (ANN). This hybridisation was performed in such a way that the function fitting algorithm learns patterns of fired rules found during the inferencing process. The position of rules and conclusions in the MCRDR structure represents the context of the knowledge, while the network adjusts its function over time as a means of capturing hidden relationships. It is these relationships that represent the methodology's hidden contexts.

This amalgamation appears simplistic but is by no means trivial. The fundamental difficulty was finding a means for taking the inferenced results from MCRDR and coding an input sequence for the network. The problem is caused by MCRDR's structure constantly expanding. Therefore, the network's input space must also grow to match. However, previous work in the function-fitting literature has not attempted to develop a network capable of increasing its input space. The problem arises from the internal structure of neural networks where, as the input space is altered, the interconnections between neurons and the associated weights are also changed.

Basically, the system discussed in this paper is designed to recognise patterns of rules for particular cases and to attach weightings to these observed patterns. These patterns exist because there is either a conscious or subconscious relationship between these classes in the expert's mind. Therefore, the captured pattern of rules in their static context is effectively a type of hidden or unknown context. This now discovered context can be given a value representing its contribution to a particular task.

1.   **Pre-process Case**
            Initialise Case *c*
            c ← Identify all useful data elements.

2.   **Classification**
            Initialize *list* to store classifications
            Loop
                    If child's rule evaluates Case c to *true*
                            list ← goto step 2 (generate all classifications in child's branch).
            Until no more children
            If no children evaluated to true then
                    list ← Add this nodes classification.
            Return *list*.

3.   **Evaluate Case**
            $\overline{x}$ ← Generate input vector from *list*.
            ANN ← $\overline{x}$
            $\overline{v}$ ← *ANN* output value.

4.   **Return RM evaluation**
            Return *list* of classifications for case *c* and
            Value $\overline{v}$ of case *c*.

Fig. 1. Pseudo code algorithm for RM

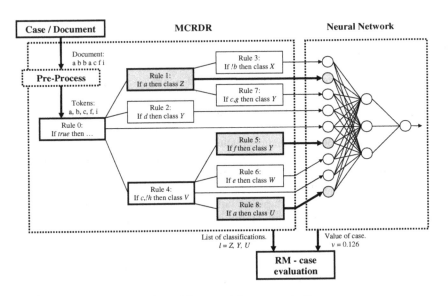

Fig. 2. RM illustrated diagrammatically

The full RM algorithm, given in pseudo-code in Fig 1 and shown diagrammatically in Fig 2, consists of two primary components. Firstly, a case is pre-processed to identify all of the usable data elements, such as stemmed words or a patient's pulse. The data elements are presented to the standard MCRDR engine, which classifies them according to the rules previously provided by the user. Secondly, for each attribute, rule or class identified, an associated input neuron in the neural network will fire. The network produces a vector of output values, $\overline{v}$, for the case presented. The system, therefore, essentially provides two separate outputs; the case's classifications and the associated set of values for those classifications.

Fig 2 shows a document classification and storage system where documents are also rated to judge their immediate importance. In this example a document with the tokens {a b b a c f i} has been pre-processed to a set of unique tokens {a, b, c, f, i}. The case is then presented to the MCRDR component of the RM system, which ripples the case down the rule tree finding three classifications: Z, Y, and U; from the terminating rules: 1, 5, and 8. In this example, which is using the Terminating Rule Association (TRA) method (section 3.3), the terminating rules then cause the three associated neurons to fire. The input pattern then feeds forward through the neural network producing a single value of 0.126. Thus, this document has been allocated a set of classifications that can be used to store the document appropriately, plus a rating indicating the importance of the document.

3.1 Learning in RM

Learning in RM is achieved in two ways. Firstly, the value for each corresponding value for receives feedback from the environment concerning its accuracy. Thus, a system using RM must provide some means of either directly gathering or indirectly estimating each elements value. For example, in an email application where the system was required to order the documents in the order of importance, the amount of reward given to the network could be based on the order the articles are read by the user or whether the user prints, saves, replies, forwards or deletes the email. How the network actually learns is either using the standard backpropagation approach using a sigmoid thresholding function, or, using the RM specific algorithm described in section 3.4. The MCRDR component still acquires knowledge in the usual way (section 2). Therefore, in the basic RM implementation the expert must still review cases and check classifications are correct.

3.2 Artificial Neural Network Component

The ANN used is based on the backpropagation algorithm and was designed to be plugged on to the end of the MCRDR component. Integration of the MCRDR and ANN components is carried out by codifying the relevant features taken from MCRDR and converting these into a single input array of values, $\overline{x}$, which is to be provided to the ANN for processing. Two methods were used in this paper referred to as the Rule Path and Terminating Rule Association methods. The rule path method provided an input for every rule that fired while the terminating method only fired input nodes associated with the final rule that was reached by the inferencing process.

3.3 Adding Neurons

As the input space grows new input nodes need to be added to the network in such a way that does not damage already learnt information. A number of methods were developed for altering the input space, such as backpropagation and radial basis function networks, as well as non neural network methods such as Kernel based methods. This paper will discuss the most stable and effective method found. This particular

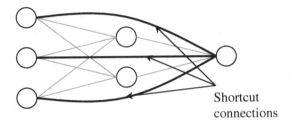

Shortcut
connections

Fig. 3. Network structure of RM showing a single hidden layer network with shortcut connections directly connecting the input nodes to the output nodes

method allows for non-linear relationships while being able to learn quickly when new inputs are added.

RM captures the initial information by directly calculating the required weight to provide us with the correct *weighted-sum* using a new learning rule referred to as the *single-step-Δ-initialisation-rule* (3.4.1). When applying this weight it must be done in so that does not affect any of the already learnt weights. Therefore, the network structure needed to be altered by adding shortcut connections (Fig 3) from any newly created input node directly to each output node and using these connections to carry the entire weight adjustment. When a new input node is added, additional hidden nodes also may be added. Therefore, connections must be added in the following places:

- From the new input node to all of the old hidden nodes.
- From all input nodes, new and old, to each of the new hidden nodes, if any.
- From each of the new hidden nodes, if any, to all of the output nodes.
- The shortcut connections from the new input node to all of the output nodes.

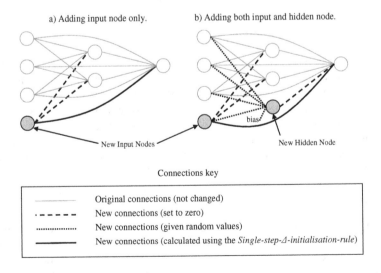

a) Adding input node only.   b) Adding both input and hidden node.

New Input Nodes   New Hidden Node

bias

Connections key

———	Original connections (not changed)
· — — —	New connections (set to zero)
··············	New connections (given random values)
———	New connections (calculated using the *Single-step-Δ-initialisation-rule*)

Fig. 4. Process used for adding new input and hidden nodes in RM. (a) shows how inputs are added by themselves. (b) shows how input and hidden nodes are added simultaneously.

The process for adding nodes and connections is illustrated in Fig 4. Each of these different groups of new connections requires particular start up values. First, the new connections from the new input node to all the old hidden nodes should be set to zero so that they have no immediate affect on current relationships. Occasionally, new hidden nodes are also required. These were added at a rate that maintained a number equivalent to half the amount of input nodes. If new hidden nodes were added then the connections from them to the output nodes should also be set to zero for the same reason as with the input nodes. In order for these connections to be trained, the output from the new hidden nodes must be non-zero. Thus, the new connections from all the input nodes to the new hidden nodes, and their biases, are given random values. Finally, the new shortcut connections are given a value calculated using the *single-step-Δ-initialisation-rule*.

3.3.1 The Single-Step-Δ-Initialisation-Rule

The *single-step-Δ-initialisation-rule* directly calculates the required weight for the network to step to the correct solution immediately. This is accomplished by reversing the feedforward process back through the inverse of the symmetric sigmoid. This calculation is performed by finding the weight needed, using equation (1), for the new input connection, w_{no}. This has the requirement that the system does not attempt to set the value of the output outside the range $-0.5 > (f(net) + δ) > 0.5$ as this will cause an error. The value for *net* for each output node, o, was previously calculated by the network during the feedforward operation where there are $n > 1$ input nodes and the nth input node is our new input. Function f is the asymmetric sigmoid and $δ$ is the amount of error. This is then divided by the input at the newly created input node, x_n, which is always 1 in this implementation, where there are $n > 0$ input nodes and $o > 0$ output nodes. Additionally, it is possible for the expert to add multiple new rules for the one case. In these situations the calculated weight is divided by the number of new features (attribute, rule or class), m. Finally, the equation is multiplied by the step-distance modifier, *Zeta (ζ)*.

$$w_{no} = ζ \left[\left(\left(\log \left[\frac{f(net)_o + δ_o + 0.5}{0.5 - (f(net)_o + δ_o)} \right] \right) / k \right) - \left[\left(\sum_{i=0}^{n-1} x_i w_{io} \right) + \left(\sum_{h=0}^{q} x_h w_{ho} \right) \right] \right] / mx_n \quad (1)$$

The *Zeta (ζ)* modifier should always be set in the range $0 ≤ ζ ≤ 1$. It is included to allow adjustments to whether a full step or partial steps should be taken for the new features. For instance, if $ζ$ is 1 then the new weights will provide a full step and any future identical cases will give the exact correct answer. A lesser value for $ζ$ causes new features to only receive a portion of their value. It was found in testing that the inclusion of the $ζ$ modifier allows better performances in some situations.

This updating method is best understood by seeing what is occurring diagrammatically. Fig 6 shows an input pattern that had a weighted sum of 3.0 at the output node. This was passed through the symmetric sigmoid function, finding the output value 0.47. The correct output after a rule was added is -0.358. The correct weight for the new input node is calculated by feeding this target back through the inverse of the sigmoid function finding the value -1.8. Therefore, the new node's weight is the difference between these two values, giving -4.8.

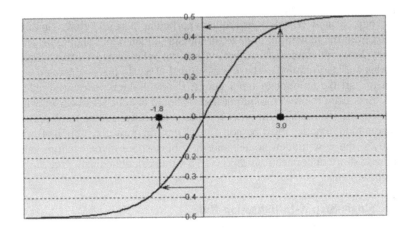

Fig. 5. Example of the *single-step-Δ-initialisation-rule* shown diagrammatically

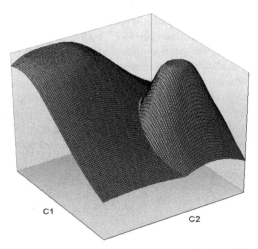

Fig. 6. Example of a randomly generated energy pattern used in the MCP simulated expert

4 Experiments and Results

This section's results are in two parts. First RM compares against the two underlying methodologies, MCRDR and a backpropagation neural network, in an online classification task. The second collection of results illustrates how RM compares against a backpropagation neural network in both on and off-line prediction. The aim of these results is to show how RM learns as fast as MCRDR, yet maintains the generalisation of a neural network. This section also contains a discussion of the simulated experts used for the experiments, along with the datasets used.

4.1 Experimental Method

In the first group of experiments the aim is to test RMs classification ability in an online environment compared to the underlying approaches. Therefore, the output from RMs ANN will consist of a vector, $\bar{v}$, of outputs. Each output will relate to one possible classification. There will be an equal number of outputs to the number of class types in the dataset being tested. If the output is positive then it will be regarded as providing that classification. The same output method is used for the backpropagation method being compared against. In the first collection of results presented in this paper each test used 10 different randomisations of the relevant datasets. The test investigates how the methods can correctly classify cases over time. In this test the entire dataset is broken up into small blocks, each $1/50^{th}$ of the original dataset, and passed through the system one group at a time. The system's performance is recorded after each group, showing how fast the system learns for each new batch of cases.

In the prediction domain RM and the ANN must output a single value, which must be as close to the expected value as possible. In the second collection of results presented in this paper each test used 10 different randomisations of the dataset. The first, *generalisation* test, divides each dataset into ten equal sized groups. Results are presented where $9/10^{ths}$ of the dataset are used for training and $1/10^{th}$ for testing. The size of the training set is then reduced incrementally in steps of $1/10^{th}$, down to $1/10^{th}$. The same $1/10^{th}$ set is always used as the test set. The *online* prediction test investigates how the methods can correctly predict values over time. Similar to the online classification test, the entire dataset is broken up into smaller blocks, each $1/50^{th}$ of the original dataset, and passed through the system one group at a time. The system's performance is recorded after each group. The value returned is then compared to the simulated expert's correct value. The absolute difference between these two values (*error*) is then averaged over all the cases in the data segment and logged.

4.2 Simulated Expertise

One of the greatest difficulties in KA and KBSs research is how to evaluate the methodologies developed [19]. The method used by the majority of RDR based research has been to build a simulated expert, from which knowledge can be acquired [19]. It is this approach that has been taken in this paper. This section will discuss the three simulated experts created for the tests performed in this paper.

4.2.1 C4.5 Simulated Expert
The only purpose of the simulated expert is to select which differences in a difference list are the primary ones. It uses its own KB to select the symbols that will make up the new KB. C4.5 [20] is used to generate the simulated expert's knowledge base. The resulting tree then classifies each case presented, just like our KB under development. If the KB being constructed, incorrectly classifies a case then the simulated expert's decision tree is used to find attributes within rules that led to the correct classification. This is similar to the *'smartest'* expert created by Compton et al [21].

4.2.2 Non-linear Multi-class Simulated Expert

The fundamental problem with the above simulated expert is that it requires an induction system, such as C4.5, to generate a complete KB prior to its use. This is a problem because no suitable system is available that can create such a tree for a multiple classification domain. However, the system being developed in this paper is primarily targeting domains with multiple classification domains. Therefore, a second simulated expert was created specifically designed for handling a particular multiple classification based dataset (4.3).

This heuristic based simulated expert has two stages in calculating classifications based on a cases attributes. The first stage uses a randomly generated table of values, representing the level that each attribute, $a \in A$, contributes to each class, $c \in C$. An example of an expert's attribute table used is shown in Table 1.

Table 1. Example of a randomly generated table used by the *non-linear multi-class* simulated expert. Attributes a - l are identified across the top, and the classes C1 – C6 down the left.

	A	b	c	d	e	f	g	h	i	j	k	l
C1	0	0	-1	3	0	0	0	0	0	0	-1	3
C2	0	0	0	-2	2	0	0	-2	0	0	1	0
C3	0	-2	1	0	0	0	0	0	0	1	0	-1
C4	-1	3	0	0	0	0	1	0	-1	0	0	0
C5	0	0	0	0	-2	2	-2	0	2	0	0	0
C6	2	0	0	0	0	-2	0	1	0	-2	0	0

To make the task sufficiently difficult for the systems to learn a second stage of the expert's classification process is to provide a non-linearity. A non-linear expert needs the attributes' contribution to classifications to vary according to what other attributes were in the case. This was achieved by selecting an even number of attributes and pairing them together for each of the classes. Once paired, they were given an increasing absolute value. Additionally, alternate pairs had their sign changed. This can best be understood by investigating the example shown in Table 2. Here it can be seen that for the class *C1*, the attribute pairs *{b, j}*, *{f, h}* and *{d, j}* have a positive influence, while *{a, l}*, *{h, i}* and *{a, k}* have a negative influence.

Table 2. Example of a randomly generated table of attribute pairs. The top numbers represent the positive or negative values for the pairs. Each class in this example has six pairs.

	1		-1		2		-2		3		-3	
C1	b	j	a	l	f	h	h	i	d	j	a	k
C2	g	b	c	f	e	h	a	b	k	d	g	k
C3	i	d	e	b	g	i	k	l	j	a	c	f
C4	l	a	c	i	j	a	i	l	f	h	j	a
C5	k	g	b	f	d	g	j	f	b	c	a	e
C6	c	l	h	j	a	c	j	b	g	k	d	e

Table 3. Two example cases being evaluated by the non-linear multi-class simulated expert

Case A = {a, b, c, d}							Case B = {a, c, e, g}						
Attributes	Classifications						Attributes	Classifications					
	1	2	3	4	5	6		1	2	3	4	5	6
a	0	0	0	-1	0	2	a	0	0	0	-1	0	2
b	0	0	-2	3	0	0	c	-1	0	1	0	0	0
c	-1	0	1	0	0	0	e	0	2	0	0	-2	0
d	3	-2	0	0	0	0	g	0	0	0	1	-2	0
{a, c}						2	{a, c}						2
{a, b}		-2					{e, a}					-3	
{b, c}				3									
Total	2	-4	-1	2	3	4	Total	-1	2	1	0	-7	4
Classified	✓	✗	✗	✓	✓	✓	Classified	✗	✓	✓	✗	✗	✓

When a case is presented to the expert the class it belongs to is calculated by adding all the attribute values and there attribute pairs. The expert will then classify the case according to which classes provided a positive, > 0, total. When creating a new rule, the expert selects the attribute or attribute-pair from the difference list that distinguishes the new case from the cornerstone case to the greatest degree. Table 3, gives two example cases where each case has 4 attributes.

4.2.3 Multi-class-Prediction Simulated Expert

Testing RM using simulation has an added difficulty in the prediction domain. This is because available datasets do not give both symbolic knowledge and a target value instead of a classification. This could be partially resolved by assigning each classification a value. However, fundamentally this would still be a classification type problem. The approach taken in this paper was to develop a third simulated expert, which has two stages in calculating a value for a case based on a set of randomly generated attributes. The first stage uses a randomly generated table of values, in the same way as the first stage of the non-linear simulated expert described above. This classification stage is merely an intermediate step to finding a rating for the case. It is also used during knowledge acquisition for identifying relevant attributes in the difference lists. When creating a new rule, the expert selects the attribute from the difference list that distinguishes the new case from the cornerstone case to the greatest degree. This was achieved by locating the most significant attribute, either positively or negatively, that appeared in the difference list (see example in Table 1).

To fully push the system's abilities, the rating calculated by the simulated expert needs to generate a non-linear value across the possible classifications. The implementation used for prediction generates an energy space across the level of class activations, giving an energy dimensionality the same as the number of classes possible. Each case is then plotted on to the energy space in order to retrieve the case's value. First, the strength of each classification found is calculated. As previously discussed a case was regarded as being a member of a class if its attribute value was greater than 0. However, no consideration was made to what was the degree of membership. In this expert the degree of the case's membership is calculated as a percentage, p, of membership using Equation 2.

$$p = t^a / t^m \qquad (2)$$

This is simply the actual calculated total, t^a, divided by the maximum possible total, t^m, for that particular class. Extending the example from Table 3 for case A, classification C1, the total 2 is divided by the best possible degree of membership 6, max value from row C1 in table 1, thereby, giving a percentage, p, membership of *33%*. This calculation is performed for each class. Each class then has a randomly selected point of highest value, or centre, c, which is subtracted from the percentage and squared, Equation 3. This provides a value which can be regarded as a distance measure, d, from the centre. This distance measure can be *stretched* or *squeezed*, widening or contracting the energy patterns around a centre, by the inclusion of a width modifier, w.

$$d = w\,(\,p - c\,)^2 \qquad (3)$$

The classes' centres are combined to represent the point of highest activation for the expert, referred to as a *peak*. Therefore, if the square root of the sum of distances is taken then the distance from this combined centre can be found. This distance can then be used to calculate a lesser value for the case's actual rating. Therefore, as a case moves away from a *peak* its *value* decreases. Any function can be used to calculate the degree of reduction in relation to distance. In this paper a Gaussian function was used. Equation 4 gives the combined function for calculating a value for each possible peak, v^p, where n is the number of classes in the dataset.

$$v^p = \frac{1}{1 + e^{-0.5\left(1 \,\middle/\, \sqrt{\sum_{j=0}^{n}\left(\left(\frac{t_j^a}{t_j^m}\right)w_j - c_j\right)^2}\right)}} - 0.5 \qquad (4)$$

Finally, it is possible to have multiple peaks in the energy space. In such a situation each class has a centre for each peak. Each peak is then calculated in the same fashion as above, resulting in a number of values, one for each peak. The expert then simply selects the highest value as the case's actual rating. This rating method is best understood by looking at a three dimensional representation shown in Fig 6.

The third dimension, shown by the height, illustrates the value at any particular point in the energy space. This figure shows a dataset with only two possible classes, C1 and C2, and two peaks. A three class dataset cannot be represented pictorially. The advantage of this approach is that it generates an energy pattern that is nonlinear. At no location can a straight line be drawn where values are all identical.

4.3 Datasets

The method was tested using six datasets. The first five are standard datasets retrieved from the University of California Irvine Data Repository [22]. These five datasets were tested using the *C4.5* based simulated expert. The sixth dataset is a purpose designed randomised set and is used with the *non-linear multi-class* and *multi-class-prediction* simulated experts. Below is a list describing each of the five dataset used from the University of California Irvine Data Repository [22].

- **Chess** – has 36 attributes with a binary classification over 3196 cases. In each 1/50th group there are 63 cases.
- **Tic-Tac-Toe (TTT)** –has 9 attributes with a binary classification over 958 cases. In each 1/50th group there are 19 cases.
- **Nursery Database** – has 8 nominal attributes with 5 classifications over 12960 cases. In each 1/50th group there are 259 cases.
- **Audiology** – has 70 nominal-valued attributes with 17 classifications over only 200 cases. In each 1/50th group there are 4 cases.
- **Car Evaluation** – has 6 attributes with 4 classifications over 1728 cases. In each 1/50th group there are 34 cases.

The multi-class dataset builds cases by randomly selecting attributes from the environment. For instance, an environment setup for the example simulated experts used in section 4.2.2 would allow for 12 possible attributes. For the tests in this paper each case selected 6 attributes, giving a possible 924 different cases. There were also 6 possible conclusions. Therefore, in each 1/10th group used in the offline prediction there are 92 cases and 18 cases in each 1/50th group.

4.4 Online Classification

One of the main features RM was aiming to achieve from the use of the ANN was the ability to learn quickly and generalise well in an online environment. The results in this section investigate how RM compares with its two underlying methodologies in the online environment. Fig 7 (a) - (f) shows how RM, MCRDR and the ANN perform on the six datasets. Each point on the charts is an average of the previous 10 data segments (except 2 data segments for the Audiology dataset) which are then further averaged over the ten randomised runs. Each segment contains a random selection of cases, each 1/50th the size of the whole dataset. Error bars have been omitted to allow for greater readability.

These comparisons are powerful indicators of the advantages of RM over a standard backpropagation neural network when being applied in an online environment. On the chess, TTT and audiology datasets it can be seen that RM has learned as fast or nearly as fast as MCRDR. On the nursery and car evaluation datasets it was only between 3% and 6% below MCRDRs performance and overtime was narrowing this gap. This meets our original goal of gaining the speed of MCRDR's instantaneous learning as soon as a rule is added. In the multi-class results this same result can be observed, except it can also be seen how RM continued to learn after MCRDR had accrued all its possible knowledge.

MCRDR's failure to continue to learn after its initial gains was a point of concern in the multi-class test. However, after investigation, it was found to be caused by two main factors. Firstly, for a case to be correctly classified it must get all six classes correct. Therefore, MCRDR's performance was not as poor as it first appears. Secondly, there is one unusual problem in the MCRDR rule creation and validation phase. It is possible that when an expert attempts to create a rule there may be no suitable attributes available. This generally only occurs on the later difference lists generated when there are multiple cornerstone cases.

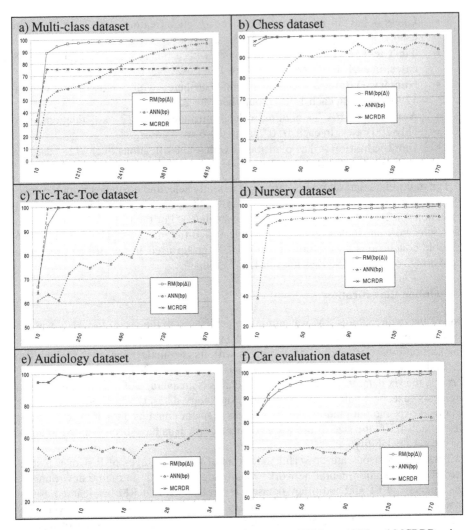

Fig. 7. (a) – (f) shows charts comparing the performance of RM, an ANN and MCRDR using different datasets. The x-axis shows the amount of 1/50th data segments that have been seen. The y-axis shows the average accuracy over the last 10 data segments.

The complexity of the multi-class dataset, especially the use of attribute pairs highlights this problem. This caused the simulated expert to be unable to create required rules on some occasions. Therefore, for the purposes of this test, failed rules were not added to the knowledge base. However, these lost rules can now be treated as a form of hidden context. Therefore, RM's ability to significantly outperform the MCRDR's performance shows its ability to capture that hidden information even when it is unavailable to the knowledge base. The performance of RM appears to essentially learn exactly like any standard learning curve but rather than start from scratch it began from where MCRDR had finished learning.

4.5 Prediction Generalisation

Traditionally, MCRDR and other KBSs can usually only be applied to classification problems. Even when used for prediction they usually still use the same basic classification style but each classification gives a predictive value instead. One advantage found with RM is that it can also be applied in a prediction environment. This can be achieved by the network being setup to output just a single value, representing the system's prediction for the task at hand.

The ability of a method to generalise is measured by how well it can correctly rate cases during testing that it did not see during training. The value returned by RM and the ANN is then compared to the simulated expert's correct value. The absolute difference between these two values (error) is then averaged over all the cases in the data segment and logged. The results shown in Fig 8 show they each performed. Each point on the charts is the average error for the test data segment averaged over ten randomised runs of the experiment, for each of the nine tests. To reduce the complexity of the charts shown, error bars have been omitted.

These results show that the RM hybrid system has done exceptionally well both initially as well as after training is complete when generalising. Additionally, it can be observed that the neural network was unable to significantly improve with more training data. This problem is caused by the network having consistently fallen into

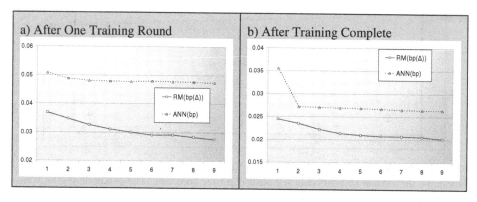

Fig. 8. (a) – (b) Two charts comparing how RM and ANN. Chart a) shows how the methods compare after only one viewing of the training set. Chart b) shows how the methods compare after training was completed. The x-axis shows how many tenths of the dataset were used for training. All results used the last tenth for testing. The y-axis shows the average error.

local minimum, a problem common to neural networks especially in prediction domains. RM is less likely to encounter this learning problem as the knowledge base provides an extra boost, similar to a momentum factor, which propels it over any local minima and closer to the true solution. Therefore, not only does RM introduce KBSs into potential applications in the prediction domain, as well as, allow for greater generalisation similar to an ANN, but it also helps solve the local minima problem.

4.6 Prediction Online

The process of RM being able to predict an accurate value in an online environment could potentially allow the use of RM in a number of environments that have previously been problematic. For instance, KBSs in *information filtering* (IF) have difficulties due to their problems in prediction, while neural networks are far too slow. RM allows for the inclusion of expert knowledge with the associated speed but also provides a means of value prediction. Fig 9 shows a comparison between RM and an ANN in an online environment. Here it can once again be observed that RM has performed outstandingly well from the outset and was able to maintain this performance. This fast initial learning can be vital in many applications as it is what users usually expect.

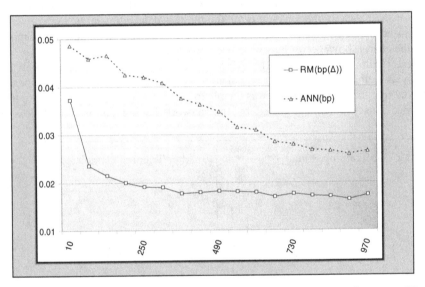

Fig. 9. This chart compares how RM and an ANN, perform in an online environment. The x-axis shows the amount of $1/50^{th}$ data segments that have been seen. The y-axis shows the average error over the last 10 data segments, also averaged over 10 trials.

5 Conclusion

This paper presented an algorithm that detects and models hidden contexts within a symbolic domain. The method developed builds on the already established Multiple Classification Ripple-Down Rules (MCRDR) approach and was referred to as Rated MCRDR (RM). RM retains a symbolic core that acts as a contextually static memory, while using a connection based approach to learn a deeper understanding of the knowledge captured.

A number of results were presented in this paper, which have shown how RM is able to acquire knowledge and learn. RM's ability to perform well can be put down to two features of the system. First, is that the flattening out of the dimensionality of the problem domain by the MCRDR component allows the system to learn a problem that is mostly linear even if the original problem domain was non-linear. This allows the network component to learn significantly faster. Second, the network gets an additional boost through the *single-step-Δ-initialisation rule*, allowing the network to start closer to the correct solution when knowledge is added.

Acknowledgements

The majority of this paper is based on research carried out while affiliated with the Smart Internet Technology Cooperative Research Centre (SITCRC) Bay 8, Suite 9/G12 Australian Technology Park Eveleigh NSW 1430 and the School of Computing, University of Tasmania, Locked Bag 100, Hobart, Tasmania.

References

1. Newell, A., Simon, H.A.: Computer Science as empirical Inquiry: Symbols and Search. Communications of the ACM 19(3), 113–126 (1976)
2. Menzies, T.: Assessing Responses to Situated Cognition. In: Tenth Knowledge Acquisition for Knowledge-Based Systems Workshop, Catelonia, Spain (1996)
3. Menzies, T.: Towards Situated Knowledge Acquisition. International Journal of Human-Computer Studies 49, 867–893 (1998)
4. Wille, R.: Restructuring Lattice Theory: An Approach Based on Hierarchies of Concepts. In: Rival, I. (ed.) Ordered Sets: Proceedings of the NATO Advanced Study Institute held at Banff, Canada, pp. 445–472. D. Reidel Publishing, Dordrecht (1981)
5. Kelly, G.A.: The Psychology of Personal Constructs, Norton, New York (1955)
6. Compton, P., Jansen, R.: Knowledge in Context: a strategy for expert system maintenance. In: Second Australian Joint Artificial Intelligence Conference (AI 1988), vol. 1, pp. 292–306 (1988)
7. Brezillon, P.: Context in Artificial Intelligence: II. Key elements of contexts. Computer and Artificial Intelligence 18(5), 425–446 (1999)
8. Dazeley, R., Kang, B.: Epistemological Approach to the Process of Practice. Journal of Mind and Machine, Springer Science+Business Media B.V. 18, 547–567 (2008)
9. Gaines, B.: Knowledge Science and Technology: Operationalizing the Enlightenment. In: Proceedings of the 6th Pacific Knowledge Acquisition Workshop, Sydney, Australia, pp. 97–124 (2000)
10. Menzies, T., Debenham, J.: Expert System Maintenance. In: Kent, A., Williams, J.G. (eds.) Encyclopaedia of Computer Science and Technology, vol. 42, pp. 35–54. Marcell Dekker Inc., New York (2000)
11. Beydoun, G.: Incremental Acquisition of Search Control Heuristics, Ph.D thesis (2000)
12. Compton, P., Edwards, G., Kang, B.: Ripple Down Rules: Possibilities and Limitations. In: 6th Banff Knowledge Acquisition for Knowledge-Based Systems Workshop (KAW 1991), vol. 1, pp. 6.1–6.18. SRDG publications, Canada (1991)
13. Compton, P., Kang, B., Preston, P.: Knowledge Acquisition Without Knowledge Analysis. In: European Knowledge Acquisition Workshop (EKAW), vol. 1, pp. 277–299. Springer, Heidelberg (1993)
14. Preston, P., Edwards, G., Compton, P.: A 1600 Rule Expert System Without Knowledge Engineers. In: Moving Towards Expert Systems Globally in the 21st Century (Proceedings of the Second World Congress on Expert Systems 1993), New York, pp. 220–228 (1993)

15. Preston, P., Edwards, G., Compton, P.: A 2000 Rule Expert System Without a Knowledge Engineer. In: Proceedings of the 8th AAAI-Sponsored Banff Knowledge Acquisition for · Knowledge-Based Systems Workshop, Banff, Canada, pp. 17.1-17.10 (1994)
16. Kang, B.: Validating Knowledge Acquisition: Multiple Classification Ripple Down Rules, Ph.D thesis (1996)
17. Kang, B.H., Compton, P., Preston, P.: Multiple Classification Ripple Down Rules: Evaluation and Possibilities. In: The 9th Knowledge Acquisition for Knowledge Based Systems Workshop. SRDG Publications, Department of Computer Science, University of Calgary, Banff, Canada (1995)
18. Preston, P., Compton, P., Edwards, G.: An Implementation of Multiple Classification Ripple Down Rules. In: Tenth Knowledge Acquisition for Knowledge-Based Systems Workshop, Department of Computer Science, University of Calgary. SRDG Publications, Calgary (1996)
19. Compton, P.: Simulating Expertise. In: Proceedings of the 6th Pacific Knowledge Acquisition Workshop, Sydney, Australia, pp. 51–70 (2000)
20. Quinlan, J.R.: C4.5: Programs for Machine Learning. Morgan Kaufmann, San Francisco (1993)
21. Compton, P., Preston, P., Kang, B.: The Use of Simulated Experts in Evaluating Knowledge Acquisition. In: 9th AAAI-sponsored Banff Knowledge Acquisition for Knowledge Base System Workshop (KAW 1995), vol. 1, pp. 12.1--12.18. SRDG publications, Canada (1995)
22. Blake, C.L., Merz, C.J.: UCI Repository of machine learning databases, University of California, Irvine, Dept. of Information and Computer Sciences (1998)

Using Formal Concept Analysis towards Cooperative E-Learning

Ghassan Beydoun

School of Information Systems and Technology, University of Wollongong, Australia
beydoun@uow.edu.au

Abstract. Student interactions in an e-learning community are captured to construct a Semantic Web (SW) to create a collective meta-knowledge structure
guiding students as they search the existing knowledge corpus. Formal Concept
Analysis (FCA) is used as a knowledge acquisition tool to process the students
virtual surfing trails to express and exploit the dependencies between web-
pages to yield subsequent and more effective focused search results. The system
KAPUST2 (*K*eeper *A*nd *P*rocessor of *U*ser *S*urfing *T*rails) which constructs
from captured students trails a conceptual lattice guiding student queries is presented. Using KAPUST as an e-learning software for an undergraduate class
over two semesters shows how the lattice evolved over the two semesters, improving its performance by exploring the relationship between 'kinds' of research assignments and the e-learning semantic web development. Course
instructors monitored the evolution of the lattice with interesting positive peda-
gogical consequences.

1 Introduction

It was proposed in [1] to use indirect social navigation [2], to exploit web-pages dependencies using surfing trails left behind. These are not beneficial if individuals surfing the net have different interests, but in a given interest group individuals produce
trails that are of interest to the whole group. Early experiments in e-learning in [1] in-
dicated processing surfing trails left by students using Formal Concept Analysis (FCA)
is possible. But lack of crossings between surfing trails lowered the usability of the
resultant lattice. Our hypothesis in this paper is that the effectiveness of the resultant
conceptual lattice depends on a sufficient complexity of the conceptual lattice itself on
one hand and the following factors on the other: the way assignments are set to ensure
trails crossings, a reflective learning setting, and a sufficient number of trails. In this
paper, the FCA based system is employed for two consecutive semesters in an American University and using comparative assignments to ensure higher number of trail
crossings. The conceptual lattice resultant from the first semester is used as a starting
point for further collective development by students in the second semester. The experiments confirm that social navigation can be simulated in e-learning settings.

To effectively use web-surfing experience of people, a user friendly exposure of
the experience that all people can understand is required. In capturing and organizing
user trails according to the browsing topic, and later allowing intelligent search of the

D. Richards and B.-H. Kang (Eds.): PKAW 2008, LNAI 5465, pp. 109–117, 2009.
© Springer-Verlag Berlin Heidelberg 2009

web, users provide their topic of interest, within their interest group, and begin browsing web pages. Submitted topics of interest and trails left behind by users form the raw information to constitute the SW structure, and determine how it evolves. Formal Concept Analysis (FCA) [3] is applied to reason about the traces instead of only displaying and browsing trails as in [4]. The reasoning component has a retrieval knowledge base (the actual SW) which integrates users knowledge scattered in their left-behind surfing traces. Our system, KAPUST, in capturing user traces, it is similar to [5] and to [4, 6] which store interactions history on a user basis.

FCA has been used in various classification tasks e.g. to classify software engineering activities [7] and to impose structure on semi-structured texts e.g. [8, 9]. It has been found efficient when applied to document retrieval systems for domain specific browsing [9]. Our use of FCA is unique in that the lattice is built in collaborative manner reflecting the collective meaning of *user trails,* and modelling *unintended* and *indirect* social navigation over the web: users are not intentionally helping each other and they do not directly communicate. We build a complex information space (the SW), where we analyze traces left by users. Our approach is similar to Footprints [4, 6], where a theory of interaction history was developed and a series of tools were built to allow navigation of history rich pages and to contextualize web pages during browsing. Unlike our approach, it does not use history to make recommendations and nor does it have a reasoning technique embedded in it.

Creating an on-line collaborative learning environment is a necessary aspect for e-learning which is the systematic use of networked information and communications technology in teaching and learning [10]. E-Learning is flexible, relatively cheap and supplies "just in time" learning opportunities. E-learning is directly underpinned by the development of and access to information and communications technology infrastructure. Creating a sense of community and understanding the on-line behaviors of the participants are also crucial [11]. E-learning techniques in the corporate world are also often used for residential workshops and staff training programs. Several efforts have been made to create e-learning environments. Notably, in George Mason University under the Program on Social and Organizational Learning (PSOL), research is being done to create and maintain a Virtual Learning Community for the participants in the program. The purpose of that research is studying the learning of the community within the developed environment and a better understanding of the dynamics of collaborative dialogue to enable more informed and sound decision making [11].

2   Using User Trails and FCA to Build a Semantic Web

A group of students sharing an assignment problem usually discuss the assignment topic meeting face to face every day at university. An interactive environment is created to simulate those discussions and turn the collective knowledge generated from such discussions into a comprehensible semantic web that can be accessed not only by the students of the current class, but also students who will take the same class in following semesters are able to access it and develop it further. Individual traces are the data points that the FCA algorithm uses for learning. Traces are stored as a sequence of URL of pages that users visit in a *browsing session* when a user from particular interest group is searching for a specialized topic. Their trails consist of sequence of

URL annotated by the session title word(s) entered at the beginning of each web session. Web page addresses and session title keywords are the building blocks for our SW. Initially entered words are checked against dictionary of existing set of keywords in the database. This minimizes the redundancy of keywords (e.g. synonyms) and corrects any syntactical errors by users. The evolved SW structure gives authenticated users recommendations in the form of categorized web page links, based on session keywords. As user trails are accumulated, browsing sessions begin to intersect one another to create new knowledge.

FCA [3] reasoning turns user traces into structured knowledge, the conceptual lattice. This involves two steps: a matrix table is constructed showing keywords that each page satisfies, a conceptual lattice is then assembled from the matrix table. FCA starts with a context $K = (G; M; I)$, where G is a set whose elements are called objects, M is a set whose elements are called attributes, and I is a binary relation between G and M [$(g;m) \in I$ is read *"object g has attribute m"*]. Formalized concepts reflect a relation between objects and attributes. A concept in the resulting conceptual lattice is formed of a set of page URLs as the extents and a set of keywords as the intents. Concepts can be a result of either a single user session or multiple sessions that intersect each other. Input to the FCA engine contains all user traces collected so far. As more traces are collected, they are incrementally fed to the FCA engine to update the existing matrix table with any new web page URLs and keywords. With every new matrix table and a set of new traces, the FCA engine reconstructs the conceptual lattice. For our learning application, the lattice is updated on weekly basis following each class. The lattice generation can be scheduled to run daily instead of weekly in case of higher usage of KAPUST. KAPUST subsequent use of the generated lattice for query management, and intelligent interface are described in the next section. Our querying algorithm takes a user's query (set of keywords) and the conceptual lattice as input. It returns as output, the web page links that best match the search query.

A lattice, L, is a tuple of two sets, (P_i, K_i), where P_i and K_i as sets of page links and keywords respectively. To illustrate query processing by KAPUST, we denote the set of potential concepts that match the user request, together with their priorities as *PotentialConcepts = {(P_i, K_i), Priority}*, where *Priority* determines how relevant the concept is to the user's query. We take it as the depth level of a concept (P_i, K_i) in the conceptual lattice in case a matching concept is found. Otherwise, we take it as a measure of how many keywords from the set of keywords entered by the user at login, *UK*, exists in a concept (P_i, K_i). The querying algorithm has the following steps:

Step 2 in the process prunes the set of keywords entered by the user by removing new keywords that do not exist in the concept lattice.

Step 3 checks for a concept that has the exact set of pruned keywords.

Step 4 handles the case where no matching concept is found. In this case, all concepts that have one or more keyword in their set of keywords that matches any keyword in *UK'* are added as potential concepts. The priority is taken as (*CountUK – CountK*) to give highest priority to the concepts that have more matching keywords. If a concept has 2 matching keywords and the set *UK'* has 3 keywords in its list, the priority will be 1, which is higher than a concept that has 1 matching keyword where the priority will be 2.

Steps 5 and 6 consider the case where a matching concept is found. They add super concepts and/or sub concepts of the matching concept. Sub concepts will have a higher priority than the super concepts because they're more specialized. The most general and most specific concepts are not considered as super or sub concepts. If no super or sub concepts are found, the matching concept itself is added to the potential list of concepts.

Step 7 orders the *potentialconcepts* for display.

Step 8 divides the category and the page links under each category then retrieved the average rating for each page link to be displayed for the user.

The strategy of choosing super and sub concepts gives the user a better perception and a wider amount of relevant information. Sub concepts contain all extents of the concept itself. This allows us to categorize the page-sets one level deeper.

3 Architecture of (KAPUST)

KAPUST architecture (Figure 1) has two components: an extensive interactive user interface and a reasoning/knowledge creating part (invisible to the user). The reasoning components implements the conceptual framework discussed in the previous section. KAPUST is a distributed intelligent system. Its deployment requires individual deployment packages on nodes of the network. This is later described after we describe the user interface. The interactive component organizes the user input to the machine learner, and interfaces the user with existing knowledge. KAPUST search results are lists of rated and categorized links, to visit any page, users may click on their links benefiting from traces of others their community. For e-learning, students get to exploit views of each other.

A web server is needed to set up and store the semantic web component of KAPUST. A database server (SQL) is also required to store traces and execute the associated FCA engine. A client machine is designated as *administrator* and accesses the FCA engine from any machine. To deploy the application, an IIS Web Server is needed to set up the website that the annotation tool, the export utility and reporting services will be using. An SQL Server is also required to install the database of the annotation tool and the FCA engine. These are needed on the server side.

In our e-learning environment, we needed students to be able to do their assignments either from home, by installing the annotation tool using or from within the university by accessing the tool using scenario 2. Due to university policies, students are not allowed to install applications on university computers. For that purpose, a citrix server was set up at the university where the annotation tool was installed. Taking into consideration that most computers at the university have citrix client installed, the users were able to connect through citrix to the server were the annotation tool was installed and use the Internet Explorer browser from that server. Another reason for using citrix server scenario is because the university computers are accessible to all students on its campus. Since we are taking a specific domain for research we don't want to mess up the data with irrelevant information that might result from the improper usage of the tool by other student who are not taking the PSPA course. Students who want to use the tool from home can also use scenario 2 if they install

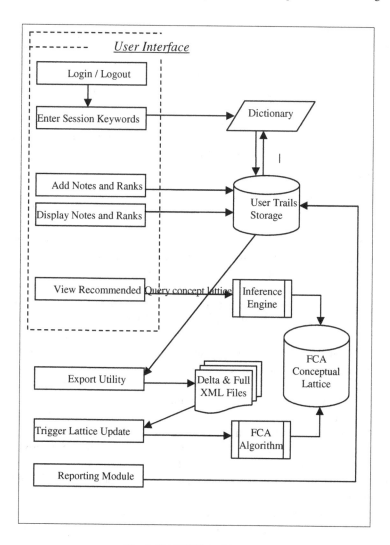

Fig. 1. KAPUST Architecture

citrix client at their PCs; but that means a slower connection and a longer waiting time compared to scenario 1 were they directly install the annotation tool at their PC. For scenario 1, a package was created that contains the annotation tool installation files. Students only have to run the package to be able to view that annotation tool as an add-on in their Internet Explorer browser and start doing their assignments.

4 E-Learning Experimental Results

For each of the first 12 weeks each semester, the professor gives a research assignment on a new topic. Browsing traces are collected on a weekly basis (as XML exported files (Delta and Full)). KAPUST also allowed the professor to track the students work and keep the collected information for another class to consume. Students are graded for their interaction with the software. Before a student searches for articles to answer the question under study, they need to log in. For each assigned question, students choose one or more keywords from the domain of the assignment question and enter them as session keywords before browsing any web pages. All visited webpages in this session are later annotated with these keywordseach assignment question is handled as a distinct browsing session by the annotation tool. At a subsequent session login the student chooses a new set of keywords representing a new question.

Semester 1 had 12 students. It was seen as an experimental setting and the presence of the professor is maintained throughout each lab. Semester1 results provided an initial conceptual structure for semester 2 when only a lab assistant was present. Students used the conceptual lattice produced by students from Semester 1 as a starting point for their learning experience and continuously added to it. Week 1 represents the testing phase where students were getting familiar with the tool. In Semester 1, weeks 2 to 7 were used to collect traces to build the semantic web structure which was built for the first time in week 8 indicating the start of knowledge sharing between students. Week 8 of semester 1 represents the maturity of the semantic web structure and the construction of the first conceptual lattice. During week 9 to 12 of semester 1, the functionality of the system as a whole is observed. In this period, the semantic web structure is continually updated online, the matrix table and the lattice structure are incrementally updated and constructed on a weekly basis, students continue to share their experiences, and most importantly students are given recommendations for each new browsing session they initiate.

The annotation tool successfully collected user traces over the 12 weeks of experimentation. A large number of traces were collected. The rate of collecting traces increased over the last few weeks. We attribute this increase to the deployment and use of the lattice structure. Using the conceptual lattice gave the students a kind of motivation to work harder and helped in their learning process. It is encouraging to get relevant links directly to the point one is searching for. Moreover, students started to cross one another's trails after week 7. The first FCA conceptual lattices were generated in week8. The inference engine started taking action during this period to query the conceptual lattice and present recommendations to the students according to their search criteria. The first full exported file is fed to the inference engine in week 8 to generate the first version of the conceptual lattice. 60 user sessions are collected, 112 notes and 156 ratings are made. Concepts are scattered across 8 levels (including most specific and general concepts). The conceptual lattice is finally formed of 255 pages, 98 keywords and 109 concepts. Table 1 displays characteristics of concepts at each of level of the final lattice. Recall that a concept is formed of a set of pages and a set of keywords (characterizing the session), level 1 contains the most general concept (see table 1). This concept is formed of all the pages in its page set and contains the empty set in its keyword set. At level 8 we have the most specialized concept. This concept contains the empty set in its page set and all the keywords in its keyword set.

Table 1. Analysis of Conceptual Lattice by Level: The number of keywords per concept increases- in reverse to number of pages, as we move towards the more specialized concepts

	# of Concepts		Keywords per concept		# of pages per concept	
	Semester 1	Semester 2	Semester 1	Semester 2	Semester 1	Semester2
Level 1 (General)	1	1	0	0	255	434
Level 2	59	121	1 to 2	1 to 3	1 to 22	1 and 23
Level 3	21	34	2 to 3	2 to 3	1 to 12	1 and 12
Level 4	16	20	3 to 5	3 to 5	1 to 9	1 and 10
Level 5	6	7	5 to 8	5 to 9	1 to 6	1 and 7
Level 6	4	5	8 to 9	7 to 8	1 to 4	1 and 6
Level 7	1	5	12	7 to 9	1	2 and 4
Level 8	1	4	98	9 to 12	0	1 and 3
Level 9	0	3	0	10 to 12	0	1 and 2
Level 10	0	1	0	12	0	1
Level 11 specific	0	1	0	173	0	0

The fact that the lattice has several levels indicates knowledge sharing by the students. It also means that the tool has been successful in mapping concepts with the corresponding pages from different user trails to generate new concepts on various levels of the lattice structure. At level 7, the second most specialized concept, we have 1 concept which refers to the *Google* search engine in its page set and which has 12 satisfied keywords. This concept is not too relevant, because the *Google* search engine is not a valid web page. It's rather a general site. This concept resulted from improper usage of the tool and from testing or noise data during week 1

Table 2. Traces and crossings during semester 2

	Week 1,2	Week 3,4	Week 5,6	Week 7,8
# of Pages Visited	23	6	24	28
# of Crossing Pages	2	2	0	4
	Week 9,10	Week 11,12	Week 13,14	
# of Pages Visited	35	32	54	
# of Crossing Pages	4	2	32	

Semester 2 had 14 students. Observations made in semester 1 led to modifying the style of questions set for semester 2 to ensure more crossings between surfing trails in the second semester. Because the already mature semantic web, it was regenerated using traces collected on a bi-weekly basis (rather than weekly). Table 2 shows the number of pages visited by users per week, as well as the number of crossing pages per week (pages that have been visited by 2 different users). Crossing pages give us a measure of how much sharing of knowledge is occurring. The number crossing pages indicates that the tool has helped students to find their target by using the stored set of previously visited pages. Compared to the results of the earlier semester, once notices that the crossing pages occurred more often and this is due to having an initial lattice structure at the beginning of the semester that further matured during the semester.

Recall that last semester, the first 7 weeks were only to build the lattice and it was not until the final weeks that crossing pages started to appear.

Table 1 shows the most general which is formed of all the pages in its page set and contains the empty set in its keyword set. At level 11 we have the most specialized concept. This concept contains the empty set in its page set and all the keywords in its keyword set. This means that there is no page that satisfies all keywords and there is no keyword that satisfies all pages visited. As in the earlier analysis, notice how the number of keywords per concept increases as we move towards the more specialized concepts, and how the number of pages decreases in that order. This shows that our lattice is appropriately constructed according to the sub-concept and super-concept definition in Formal Concept Analysis. More importantly, the lattice formed in semester 2 was clearly deeper. Semester 1 final lattice had 109 concepts scattered over 8 levels. Semester 2 lattice continued to grow and had increased by 3 levels with 93 additional concepts.

In our e-learning domain, traces were collected on weekly basis and the regeneration process was ran every week or bi-weekly basis and as a batch process offline. However, if the system is to be deployed in another domain where lattices are developed many times per hour, then our FCA algorithm may need to be reconsidered.

5 Discussion and Conclusion

Formal Concept Analysis generated conceptual structures provide to the students a user friendly natural presentation of the semantic web evolved. The conceptual lattice structures the data from the most general to the most specific concept. It relates concepts to each other based on their intents and extents, thus providing a means for creating new data that was not directly perceived from the user trails. Moreover, querying the lattice is an easy task and it can vary between approaches to make the most out of the structure. For instance, in our method, if a user is searching for a certain concept, we provide him with a categorized result formed of the upper and lower levels of the concept itself in order to gain more insights about the extents constituting the concept.

From KAPUST perspective, a student creates a knowledge in reaction to the search context the student is in. This context is firstly defined by at least the state of the knowledge base, the assignment and the student him/herself and secondly by the webpage visited. A knowledge base developed by the body of students will be conceptual description of the hypertext body. In other words, the students are guided to impose structure on the hypertext body. This is suggested as suitable web based activity by [12]. This also suggests that KAPUST is a tool which is not suitable for problem based subjects rather more to subjects requiring synthesis of large body of information where the lecturer is able to design assignments and unleash students as a group on the text body using KAPUST. This is a common distinction for university courses used by [12, 13] amongst others.

The non-intrusive nature of our approach solves the problem of getting feedback from customers who may not have any true interest or inclination to give it. This may be added to KAPUST by using incremental techniques described in [14][15]. We are in the process of integrating both of those mechanisms from [16] to identify more prominent evolutionary factors, and devising a measure of trust from [17] to evaluate the reliability of the outcome knowledge.

References

1. Beydoun, G., Kultchitsky, R., Manasseh, G.: Evolving Semantic Web with Social Naviga-tion. Expert Systems with Applications 32, 265–276 (2007)
2. Dieberger, A.: Supporting Social Navigation on the World Wide Web. Internation Journal of Human Computer Studies 46(6), 805–825 (1997)
3. Ganter, B., Wille, R.: Formal Concept Analysis: Mathematical Foundations. Springer, Heidelberg (1999)
4. Wexelblat, A.: History-Based Tools for Navigation. In: IEEE's 32nd Hawai'i International Conference on System Sciences (HICSS 1999). IEEE Computer Society Press, Hawai (1999)
5. Laus, F.O.: Tracing User Interactions on World-Wide Webpages, in Psychologisches Insti-tut III, Westfälische Wilhelms-Universität: Germany (2001)
6. Wexelblat, A., Maes, P.: Footprints: History-Rich Tools for Information Foraging. In: Conference on Human Factors in Computing Systems (CHI 1999), Pittsburgh (1999)
7. Tilley, T., et al.: A Survey of Formal Concept Anlaysis Support for Software Engineering Activities. In: Ganter, B., Stumme, G., Wille, R., et al. (eds.) Formal Concept Analysis. LNCS, vol. 3626, pp. 250–271. Springer, Heidelberg (2005)
8. Eklund, P., Wormuth, B.: Restructuring Help Systems Using Formal Concept Analysis. In: Ganter, B., Godin, R. (eds.) ICFCA 2005. LNCS, vol. 3403, pp. 129–144. Springer, Heidelberg (2005)
9. Kim, M.H., Compton, P.: Formal Concept Analysis for Domain-Specific Document Re-trieval Systems. In: 14th Biennial Conference of the Canadian Society for Computational Studies of Intelligence (AI 2001). Springer, Ottawa (2001)
10. Horton, W., Horton, K.: E-learning Tools and Technologies. John Wiley and Sons, Chich-ester (2003)
11. Blunt, R., Ahearn, C.: Creating a Virtual Learning Community. In: The Sixth International Conference on Asynchronous Learning Networks (ALN 2000), Maryland (2000)
12. Barker, P.: Technology in support of learning. In: Baillie, C., Moore, I. (eds.) Effetcive Learning and Teaching in Engineering. Taylor and Francis Group, Abington (2004)
13. Biggs, J.: Teaching for quality learning at university. Open University Press, Buckingham (1999)
14. Esteva, M.: Electronic Institutions: From Specification To Development, in Artificial Intel-ligence Research Insitute. UAB - Universitat Autònma de Barcelona, Barcelona (2003)
15. Drake, B., Beydoun, G.: Using Ripple Down Rules to Build Predicates in Situation Se-mantics. In: 6th Pacific Rim Knowledge Acquisition Workshop (PKAW 2000). University of New South Wales, Sydney (2000)
16. Beydoun, G., Debenham, J., Hoffmann, A.: Integrating Agents Roles Using Messaging Structure. In: Pacific Rim Multi Agent System Workshop (PRIMA 2004), Auckland Uni-versity, Auckland (2004)
17. Beydoun, G., et al.: Cooperative Modeling Evaluated. International Journal of Cooperative Information Systems 14(1), 45–71 (2005)

Navigation and Annotation with Formal Concept Analysis

(Extended Abstract)

Peter Eklund[1] and Jon Ducrou[2]

[1] School of Information Systems and Technology
University of Wollongong, NSW 2522, Australia
peklund@uow.edu.au
[2] Amazon.com Inc., Seattle, WA
jonducrou@gmail.com

Abstract. AnnotationSleuth is software to experiment with knowledge acquisition using Formal Concept Analysis (FCA). The system extends a standard search and browsing interface to feature a conceptual neighborhood centered on a formal concept derived from curatorial tags in a museum management system. The neighborhood of a concept is decorated with upper and lower neighbors in a concept lattice. Movement through the information space is achieved by the addition and removal of attributes resulting in more general and specialized formal concepts respectively. AnnotationSleuth employs several search methods: attribute search based on a control vocabulary, query refinement and query-by-example. A number of management interfaces are included to enable content to be added and tagged, the control vocabulary to be extended or conceptual scales to be defined. AnnotationSleuth, is an extensible environment for the creation of attribute lists and conceptual scales and can subsequently be used to flexibly navigate a collection of digital objects based on any user-defined semantic theme.

The Virtual Museum of the South Pacific is a collaborative project to create a proof-of-concept knowledge management system for 400 objects from the Australian Museum's (AusMus) Pacific Island collections. This project is devised to test a new means of facilitating access for Indigenous people and researchers to museum-based digital collections whose artefacts are physically distributed and often not on public display. The project has two dimensions: at the technical level the focus is on leveraging metadata used in curatorial management to produce an information system for representing collection resources as a structured associative network; at a museological level the focus is on studying the effective means of presenting and interacting with this network for traditional owners, the general public, researchers and curators. The idea of a virtual museum of the South Pacific allows more items from a catalogue to be publicly displayed than would be possible within the space constraints of a physical museum.

This research tests the relevance of concept-lattice based searching and browsing in the context of organising digitized museum content – specifically in terms

D. Richards and B.-H. Kang (Eds.): PKAW 2008, LNAI 5465, pp. 118–121, 2009.

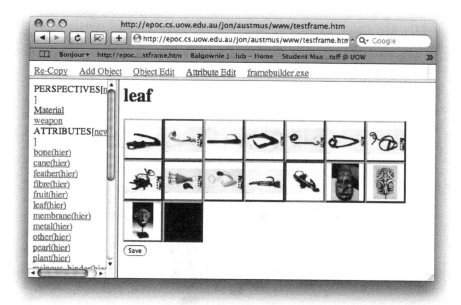

Fig. 1. Adding or editing an object's attributes. In this figure all the objects highlighted (red - 2nd, 9th and 12th) have the attribute leaf.

of enabling contextually sensitive annotation and access to the AusMus Pacific Island Collections. The project investigates the concept-lattice's potential for multi-dimensional browsing of virtual museum collection resources, developing its under-researched capacities (mostly refinement of the interface issues) and as a device to navigate and annotate sets of museum objects.

In terms of deployment, the virtual museum of the South Pacific is in its early phase but the technology on which it is based is quite mature. This paper is intended to accompany a keynote presentation at PKAW2008 on Annotation-Sleuth for an audience interested in the practical application of formal concept lattices for knowledge acquisition. Some sample images from the AusMus Vanuatu collection are used in this paper to illustrate the AnnotationSleuth.

Literature and Survey Background: Kim and Compton [5] developed *KANavigator* using FCA and a conceptual neighborhood display. Their program uses annotated documents that can be browsed by keyword. It displays the direct neighborhood (in particular the lower neighbors) in its interface. At the time, *KANavigator*'s emphasis on the use of textual labels as representations of single formal concepts (as opposed to a line diagram of the concept lattice) was a departure from FCA traditions. However, subsequent studies have shown that such an interface has significant usability merit [4], simplifying the interaction and enabling non-expert users to interact with a concept lattice.

ImageSleuth [3] has a similar interface design to explore image collections. By displaying upper and lower neighbors of the current concept and allowing navigation to these concepts, users revise their relative position in the information

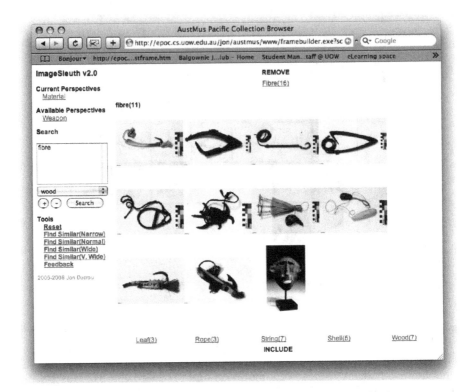

Fig. 2. AnnotationSleuth: the browsing interface. With the material perspective with the attribute fibre as a filter. Note the lower attributes Leaf, Rope, String, Shell and Wood decorated with the size of their extents.

space. This is aided by pre-defined conceptual scales that can be combined to define the attribute set of the lattice. ImageSleuth underwent iterative design and usability testing methodology reported in [2]. In addition to established methods such as keyword search and inclusion of attributes that move the user to upper/lower neighbors, a query-by-example function and restrictions to attribute set are included. Metrics on formal concept similarity are discussed elsewhere [1] and applied to ranking and automated discovery of relevant concepts: from both concepts and semi-concepts.

AnnotationSleuth follows from ImageSleuth and employs the same conceptual neighbourhood paradigm for browsing and display purposes. The program is supplemented with the ability to tag and annotated objects (images and their metadata) using a control vocabulary and a wiki. Further, a set of management interfaces allows users to create "perspectives" which correspond to FCA notion of concept scales. The user can edit and add new attributes that can then be used to create further perspectives. New objects can also be added to the system allowing the virtual collection content to grow. The program therefore represents an extensible museum content management system with a flexible mechanism for browsing and exploring the set of objects.

Browsing the Collection: AnnotationSleuth uses most of its interface to show thumbnails of images in the extent of the chosen concept (see Fig. 2). The lattice structure around the current concept is represented through the list of upper and lower neighbours which allow the user to move to super- or sub-concepts. For every upper neighbour (C, D) of the current concept (A, B) the user is allowed to remove the set $B \setminus D$ of attributes from the current intent. Dually, for every lower neighbour (E, F) the user may include the set $F \setminus B$ of attributes which takes them to a lower neighbour. By offering the sets $B \setminus D$ and $F \setminus B$ dependencies between these attributes are shown. Moving to the next concept not having a chosen attribute in its intent may imply the removal of a whole set of attributes. In order to ensure that the extent of the given concept is never empty it ios not possible to move to the bottom-most concept if its extent is empty.

Adding Narratives to the Collection: Management interfaces that allows AnnotationSleuth to be used extensibly by its user communities is an important design principle. Our project identifies 4 groups that will require different levels of access control to the virtual collection: (i) museum curators (ii) anthropologists (and other researchers such as historians) (iii) indigenous communities (including the traditional owners of the museum artifacts) and (iv) the general public.

Integrating a wiki with AnnotationSleuth is the first step towards providing user communities – the general public and also indigenous owners – with the ability to add and enhance the virtual collection. The wiki provides some aspects of access control and versioning that will be important to the final Annotation-Sleuth system. Just as in other wiki's the content added will need to be checked and moderated and users will be asked to register and verify their identity by email. A lightweight workflow engine using email to the curatorial staff will be incorporated so that comments and narratives added to objects can be checked and moderated by the museum curators. The other management interfaces, for adding objects attributes and perspectives, will have more restricted access.

References

1. Dau, F., Ducrou, J., Eklund, P.: Concept similarity and related categories in search-sleuth. In: Eklund, P., Haemmerlé, O. (eds.) ICCS 2008. LNCS, vol. 5113, pp. 255–268. Springer, Heidelberg (2008)
2. Ducrou, J., Eklund, P.: Faceted document navigation using conceptual structures. In: Hitzler, P., Schärf, H. (eds.) Conceptual Structures in Practice, pp. 251–278. CRC Press, Boca Raton (2008)
3. Ducrou, J., Vormbrock, B., Eklund, P.: FCA-based Browsing and Searching of a Collection of Images. In: Schärfe, H., Hitzler, P., Øhrstrøm, P. (eds.) ICCS 2006. LNCS (LNAI), vol. 4068, pp. 203–214. Springer, Heidelberg (2006)
4. Eklund, P., Ducrou, J., Brawn, P.: Concept lattices for information visualization: Can novices read line diagrams. In: Eklund, P. (ed.) ICFCA 2004. LNCS (LNAI), vol. 2961, pp. 57–73. Springer, Heidelberg (2004)
5. Kim, M., Compton, P.: The perceived utility of standard ontologies in document management for specialized domains. International Journal of Human-Computer Studies 64(1), 15–26 (2006)

What Does an Information Diffusion Model Tell about Social Network Structure?

Takayasu Fushimi[1], Takashi Kawazoe[1], Kazumi Saito[1], Masahiro Kimura[2], and Hiroshi Motoda[3]

[1] School of Administration and Informatics, University of Shizuoka
52-1 Yada, Suruga-ku, Shizuoka 422-8526, Japan
`k-saito@u-shizuoka-ken.ac.jp`
[2] Department of Electronics and Informatics, Ryukoku University
Otsu, Shiga 520-2194, Japan
`kimura@rins.ryukoku.ac.jp`
[3] Institute of Scientific and Industrial Research, Osaka University
8-1 Mihogaoka, Ibaraki, Osaka 567-0047, Japan
`motoda@ar.sanken.osaka-u.ac.jp`

Abstract. In this paper, we attempt to answer a question "What does an information diffusion model tell about social network structure?" To this end, we propose a new scheme for empirical study to explore the behavioral characteristics of representative information diffusion models such as the IC (Independent Cascade) model and the LT (Linear Threshold) model on large networks with different community structure. To change community structure, we first construct a GR (Generalized Random) network from an originally observed network. Here GR networks are constructed just by randomly rewiring links of the original network without changing the degree of each node. Then we plot the expected number of influenced nodes based on an information diffusion model with respect to the degree of each information source node. Using large real networks, we empirically found that our proposal scheme uncovered a number of new insights. Most importantly, we show that community structure more strongly affects information diffusion processes of the IC model than those of the LT model. Moreover, by visualizing these networks, we give some evidence that our claims are reasonable.

1 Introduction

We can now obtain digital traces of human social interaction with some relating topics in a wide variety of on-line settings, like Blog (Weblog) communications, email exchanges and so on. Such social interaction can be naturally represented as a large-scale social network, where nodes (vertices) correspond to people or some social entities, and links (edges) correspond to social interaction between them. Clearly these social networks reflect complex social structures and distributed social trends. Thus, it seems worth putting some effort in attempting to find empirical regularities and develop explanatory accounts of basic functions in the social networks. Such attempts would be valuable for understanding

D. Richards and B.-H. Kang (Eds.): PKAW 2008, LNAI 5465, pp. 122–136, 2009.

social structures and trends, and inspiring us to lead to the discovery of new knowledge and insights underlying social interaction.

A social network can also play an important role as a medium for the spread of various information [7]. For example, innovation, hot topics and even malicious rumors can propagate through social networks among individuals, and computer viruses can diffuse through email networks. Previous work addressed the problem of tracking the propagation patterns of topics through network spaces [3,1], and studied effective "vaccination" strategies for preventing the spread of computer viruses through networks [8,2]. Widely-used fundamental probabilistic models of information diffusion through networks are the *independent cascade (IC) model* and the *linear threshold (LT) model* [4,3]. Researchers have recently investigated the problem of finding a limited number of influential nodes that are effective for the spread of information through a network under these models [4,5]. Moreover, the influence maximization problem has recently been extended to general influence control problems such as a contamination minimization problem [6].

To deepen our understanding of social networks and accelerating study on information diffusion models, we attempt to answer a question "What does an information diffusion model tell about social network structure?" We except that such attempts derive some improved methods for solving a number of problems based on information diffusion models such as the influence maximization problem [5]. In this paper, we propose a new scheme for emperical study to explore the behavioral characteristics of representative information diffusion models such as the IC model and the LT model on large networks with different community structure. We perform extensive numerical experiments on two large real networks, one generated from a large connected trackback network of blog data, resulting in a directed graph of $12,047$ nodes and $79,920$ links, and the other, a network of people, generated from a list of people within a Japanese Wikipedia, resulting in an undirected graph of $9,481$ nodes and $245,044$ links. Through these experiments, we show that our proposed scheme could uncover a number of new insights on information diffusion processes of the IC model and the LT model.

2   Information Diffusion Models

We mathematically model the spread of information through a directed network $G = (V, E)$ under the IC or LT model, where V and E ($\subset V \times V$) stands for the sets of all the nodes and links, respectively. We call nodes *active* if they have been influenced with the information. In these models, the diffusion process unfolds in discrete time-steps $t \geq 0$, and it is assumed that nodes can switch their states only from inactive to active, but not from active to inactive. Given an initial set S of active nodes, we assume that the nodes in S have first become active at time-step 0, and all the other nodes are inactive at time-step 0.

2.1   Independent Cascade Model

We define the IC model. In this model, for each directed link (u, v), we specify a real value $\beta_{u,v}$ with $0 < \beta_{u,v} < 1$ in advance. Here $\beta_{u,v}$ is referred to as the *propagation probability* through link (u, v). The diffusion process proceeds from a given initial active set S in the following way. When a node u first becomes active at time-step t, it is given a single chance to activate each currently inactive child node v, and succeeds with probability $\beta_{u,v}$. If u succeeds, then v will become active at time-step $t + 1$. If multiple parent nodes of v first become active at time-step t, then their activation attempts are sequenced in an arbitrary order, but all performed at time-step t. Whether or not u succeeds, it cannot make any further attempts to activate v in subsequent rounds. The process terminates if no more activations are possible.

For an initial active set S, let $\varphi(S)$ denote the number of active nodes at the end of the random process for the IC model. Note that $\varphi(S)$ is a random variable. Let $\sigma(S)$ denote the expected value of $\varphi(S)$. We call $\sigma(S)$ the *influence degree* of S.

2.2   Linear Threshold Model

We define the LT model. In this model, for every node $v \in V$, we specify, in advance, a *weight* $\omega_{u,v} (> 0)$ from its parent node u such that

$$\sum_{u \in \Gamma(v)} \omega_{u,v} \leq 1,$$

where $\Gamma(v) = \{u \in V; (u, v) \in E\}$. The diffusion process from a given initial active set S proceeds according to the following randomized rule. First, for any node $v \in V$, a *threshold* θ_v is chosen uniformly at random from the interval $[0, 1]$. At time-step t, an inactive node v is influenced by each of its active parent nodes, u, according to weight $\omega_{u,v}$. If the total weight from active parent nodes of v is at least threshold θ_v, that is,

$$\sum_{u \in \Gamma_t(v)} \omega_{u,v} \geq \theta_v,$$

then v will become active at time-step $t + 1$. Here, $\Gamma_t(v)$ stands for the set of all the parent nodes of v that are active at time-step t. The process terminates if no more activations are possible.

The LT model is also a probabilistic model associated with the uniform distribution on $[0, 1]^{|V|}$. Similarly to the IC model, we define a random variable $\varphi(S)$ and its expected value $\sigma(S)$ for the LT model.

2.3   Bond Percolation Method

First, we revisit the bond percolation method [5]. Here, we consider estimating the influence degrees $\{\sigma(v; G); v \in V\}$ for the IC model with propagation probability p in graph $G = (V, E)$. For simplicity we assigned a uniform value p for $\beta_{u,v}$.

It is known that the IC model is equivalent to the bond percolation process that independently declares every link of G to be "occupied" with probability p [7].

It is known that the LT model is equivalent to the following bond percolation process [4]: For any $v \in V$, we pick at most one of the incoming links to v by selecting link (u, v) with probability $\omega_{u,v}$ and selecting no link with probability $1 - \sum_{u \in \Gamma(v)} \omega_{u,v}$. Then, we declare the picked links to be "occupied" and the other links to be "unoccupied". Note here that the equivalent bond percolation process for the LT model is considerably different from that of IC model.

Let M be a sufficiently large positive integer. We perform the bond percolation process M times, and sample a set of M graphs constructed by the occupied links,

$$\{G^m = (V, E^m); \; m = 1, \cdots, M\}.$$

Then, we can approximate the influence degree $\sigma(v; G)$ by

$$\sigma(v; G) \simeq \frac{1}{M} \sum_{m=1}^{M} |\mathcal{F}(v; G^m)|.$$

Here, for any directed graph $\tilde{G} = (V, \tilde{E})$, $\mathcal{F}(v; \tilde{G})$ denotes the set of all the nodes that are *reachable* from node v in the graph. We say that node u is reachable from node v if there is a path from u to v along the links in the graph. Let

$$V = \bigcup_{u \in \mathcal{U}(G^m)} \mathcal{S}(u; G^m)$$

be the strongly connected component (SCC) decomposition of graph G^m, where $\mathcal{S}(u; G^m)$ denotes the SCC of G^m that contains node u, and $\mathcal{U}(G^m)$ stands for a set of all the representative nodes for the SCCs of G^m. The bond percolation method performs the SCC decomposition of each G^m, and estimates all the influence degrees $\{\sigma(v; G); v \in V\}$ in G as follows:

$$\sigma(v; G) = \frac{1}{M} \sum_{m=1}^{M} |\mathcal{F}(u; G^m)|, \quad (v \in \mathcal{S}(u; G^m)), \tag{1}$$

where $u \in \mathcal{U}(G^m)$.

3   Proposed Scheme for Experimental Study

We technically describe our proposed scheme for empirical study to explore the behavioral characteristics of representative information diffusion models on large networks different community structure. In addition, we present a method for visualizing such networks in terms of community structure. Hereafter, the degree of a node v, denoted by $deg(v)$, means the number of links connecting from or to the node v.

3.1   Affection of Community Structure

As mentioned earlier, our scheme consists of two parts. Namely, to change community structure, we first construct a GR (generalized random) network from an originally observed network. Here GR networks are constructed just by randomly rewiring links of the original network without changing the degree of each node [7]. Then we plot the influence degree based on an information diffusion model with respect to the degree of each information source node.

First we describe the method for constructing a GR network. By arbitrary ordering all links in a given original network, we can prepare a link list $L_E = (e_1, \cdots, e_{|E|})$. Recall that each directed link consists of an ordered pair of *from*-part and *to*-part nodes, i.e., $e = (u, v)$. Thus, we can produce two node lists from the list L_E, that is, the *from*-part node list L_F and the *to*-part node list L_T. Clearly the frequency of each node v appearing in L_F (or L_T) is equivalent to the out (or in) degree of the node v. Therefore, by randomly reordering the node list L_T, then concatenating it with the other node list L_F, we can produce a link list for a GR network. More specifically, let L'_T be a shuffled node list, and we denote the i-th order element of a list L by $L(i)$, then the link list of the GR network is $L'_E = ((L_F(1), L'_T(1)), \cdots, (L_F(|E|), L'_T(|E|)))$. Here note that to fairly compare the GR network with original one in terms of influence degree, we excluded some types of shuffled node lists, each of which produces a GR network with self-links of some node or multiple-links between any two nodes.

By using the bond percolation method described in the previous section, we can efficiently obtain the influence degree $\sigma(v)$ for each node v. Thus we can straightforwardly plot each pair of $deg(v)$ and $\sigma(v)$. Moreover, to examine their tendency of nodes with the same degree δ, we also plot the average influence degree $\mu(\delta)$ calculated by

$$\mu(\delta) = \frac{1}{|\{v : deg(v) = \delta\}|} \sum_{\{v:deg(v)=\delta\}} \sigma(v). \tag{2}$$

Clearly we can guess that nodes with larger degrees influence many other nodes in any information diffusion models, but we consider that it is worth examining its curves in more details.

3.2 Visualization of Community Structure

In order to intuitively grasp the original and GR networks in terms of community structure, we present a visualization method that is based on the cross-entropy algorithm [11] for network embedding, and the k-core notion [10] for label assignment.

First we describe the network embedding problem. Let $\{\mathbf{x}_v : v \in V\}$ be the embedding positions of the corresponding $|V|$ nodes in an R dimensional Euclidean space. As usual, we define the Euclidean distance between $\mathbf{x}_u$ and $\mathbf{x}_w$ as follows:

$$d_{u,w} = \|\mathbf{x}_u - \mathbf{x}_w\|^2 = \sum_{r=1}^{R} (x_{u,r} - x_{w,l})^2.$$

Here we introduce a monotonic decreasing function $\rho(s) \in [0, 1]$ with respect to $s \geq 0$, where $\rho(0) = 1$ and $\rho(\infty) = 0$. Let $a_{u,w} \in \{0, 1\}$ be an adjacency information between two nodes u and w, indicating whether their exist a link between them ($a_{u,w} = 1$) or not ($a_{u,w} = 0$). Then we can introduce a cross-entropy (cost) function between $a_{u,w}$ and $\rho(d_{u,w})$ as follows:

$$\mathcal{E}_{u,w} = -a_{u,w}\ln\rho(d_{u,w}) - (1 - a_{u,w})\ln(1 - \rho(d_{v,w})).$$

Since $\mathcal{E}_{u,w}$ is minimized when $\rho(d_{u,w}) = a_{u,w}$, this minimization with respect to $\mathbf{x}_u$ and $\mathbf{x}_w$ basically coincides with our problem setting. In this paper, we employ $\rho(s) = \exp(-s/2)$ as the monotonic decreasing function. Then the total cost function (objective function) can be defined as follows:

$$\mathcal{E} = \frac{1}{2}\sum_{u\in V}\sum_{w\in V} a_{u,w}d_{u,w} - \sum_{u\in V}\sum_{w\in V}(1 - a_{u,w})\ln(1 - \rho(d_{u,w})). \qquad (3)$$

Namely the cross-entropy algorithm minimizes the objective function defined in (3) with respect to $\{\mathbf{x}_v : v \in V\}$.

Next we explain the k-core notion. For a given node v in the network $G = (V_G, E_G)$, we denote $A_G(v)$ as a set of *adjacent nodes* of v as follows:

$$A_G(v) = \{w : \{v, w\} \in E_G\} \cup \{u : \{u, v\} \in E_G\}.$$

A subnetwork $C(k)$ of G is called k-*core* if each node in $C(k)$ has more than or equal to k adjacent nodes in $C(k)$. More specifically, we can define k-core subnetwork as follows. For a given order k, the k-core is a subnetwork $C(k) = (V_{C(k)}, E_{C(k)})$ consisting of the following node set $V_{C(k)} \subset V_G$ and link set $V_{C(k)} \subset V_G$:

$$V_{C(k)} = \{v : |A_{C(k)}(v)| \geq k\}, \quad E_{C(k)} = \{e : e \subset V_{C(k)}\}.$$

Here according to our purpose, we focus on the subnetwork of maximum size with this property as a k-core subnetwork $C(k)$.

Finally we describe the label assignment strategy. As a rough necessary condition, we assume that each community over a network includes a higher order k-core as its part. Here we consider that a candidate for such higher core order is greater than the average degree calculated by $\bar{d} = |E|/|V|$. Then we can summarize our visualization method as follows: after embedding a given network into an R (typically $R = 2$) dimensional Euclidean space by use of the cross-entropy algorithm, our visualization method plots each node position by changing the appearance of nodes belonging to its $(\lfloor\bar{d}\rfloor + 1)$-core subnetwork. Here note that $\lfloor\bar{d}\rfloor$ denotes the greatest integer smaller than $\bar{d}$. By this visualization method, we can expect to roughly grasp community structure of a given network.

4 Experimental Evaluation

4.1 Network Data

In our experiments, we employed two sets of real networks used in [5], which exhibit many of the key features of social networks as shown later. We describe the details of these network data.

The first one is a trackback network of blogs. Blogs are personal on-line diaries managed by easy-to-use software packages, and have rapidly spread through the World Wide Web [3]. Bloggers (*i.e.*, blog authors) discuss various topics by using trackbacks. Thus, a piece of information can propagate from one blogger to another blogger through a trackback. We exploited the blog "Theme salon of blogs" in the site "goo"[1], where a blogger can recruit trackbacks of other bloggers by registering an interesting theme. By tracing up to ten steps back in the trackbacks from the blog of the theme "JR Fukuchiyama Line Derailment Collision", we collected a large connected trackback network in May, 2005. The resulting network had 12,047 nodes and 79,920 directed links, which features the so-called "power-law" distributions for the out-degree and in-degree that most real large networks exhibit. We refer to this network data as the blog network.

The second one is a network of people that was derived from the "list of people" within Japanese Wikipedia. Specifically, we extracted the maximal connected component of the undirected graph obtained by linking two people in the "list of people" if they co-occur in six or more Wikipedia pages. The undirected graph is represented by an equivalent directed graph by regarding undirected links as bidirectional ones[2]. The resulting network had 9,481 nodes and 245,044 directed links. We refer to this network data as the Wikipedia network.

4.2   Characteristics of Network Data

Newman and Park [9] observed that social networks represented as undirected graphs generally have the following two statistical properties that are different from non-social networks. First, they show positive correlations between the degrees of adjacent nodes. Second, they have much higher values of the *clustering coefficient* C than the corresponding *configuration model* defined as the ensemble of GR networks. Here, the clustering coefficient C for an undirected network is defined by

$$C = \frac{1}{|V|} \sum_{u \in V} \frac{|\{(v \in V, w \in V) : v \neq w, w \in A_G(v)\}|}{|A_G(u)|(|A_G(u)| - 1)}.$$

Another widely-used statistical measure of network is the average length of shortest paths between any two nodes defined by

$$L = \frac{1}{|V|(|V| - 1)} \sum_{u \neq v} l(u, v).$$

where $l(u, v)$ denotes the shortest path length between nodes u and v. In terms of information diffusion processes, when L becomes smaller the probability that any information source nodes can activate the other nodes, becomes larger in general.

---

[1] http://blog.goo.ne.jp/usertheme/
[2] For simplicity, we call a graph with bi-directional links an undirected graph.

Table 1. Basic statistics of networks

network	$|V|$	$|E|$	C	L
original blog	12,047	79,920	0.26197	8.17456
GR blog	12,047	79,920	0.00523	4.24140
original Wikipedia	9,481	245,044	0.55182	4.69761
GR Wikipedea	9,481	245,044	0.04061	3.12848

Table 1 shows the basic statistics of the blog and Wikipedia networks, together with their GR networks. We can see that the measured value of C for the original blog network is substantially larger than that of the GR blog network, and the measured value of L for the original blog network is somehow larger than that of the GR blog network indicating that there exisit communities. We can observe a similar tendency for the Wikipedia networks. Note that we have already confirmed for the original Wikipedia network that the degrees of adjacent nodes were positively correlated, although we derived the network from Japanese Wikipedia. Therefore, we can say that the Wikipedia network has the key features of social networks.

4.3   Experimental Settings

We describe our experimental settings of the IC and LT models. In the IC model, we assigned a uniform probability β to the propagation probability $\beta_{u,v}$ for any directed link (u,v) of a network, that is, $\beta_{u,v} = \beta$. As our β setting, we employed a reciprocal of the average degree, i.e., $\beta = |V|/|E|$. The resulting propagation probability for the original and GR blog networks was $\beta = 0.1507$, and $\beta = 0.0387$ for the original and GR Wikipedia networks. Incidentally, these values were reasonably close to those used in former study, i.e., $\beta = 0.2$ for the blog networks and $\beta = 0.03$ for the Wikipedia networks were used in the former experiments [6].

In the LT model, we uniformly set weights as follows. For any node v of a network, the weight $\omega_{u,v}$ from a parent node $u \in \Gamma(v)$ is given by $\omega_{u,v} = 1/|\Gamma(v)|$. This experimental setting is exactly the same as the one performed in [5].

For the proposed method, we need to specify the number M of performing the bond percolation process. In the experiments, we used $M = 10,000$ [5]. Recall that the parameter M represents the number of bond percolation processes for estimating the influence degree $\sigma(v)$ of a given initial active node v. In our preliminary experiments, we have already confirmed that the influence degree of each node for these networks with $M = 10,000$ are comparable to those with $M = 300,000$.

4.4   Experimental Results Using Blog Network

Figure 1a shows the influence degree based on the IC model with respect to the degree of each information source node over the original blog network, Figure 1b shows those of the IC model over the GR blog network, Figure 1c shows those of

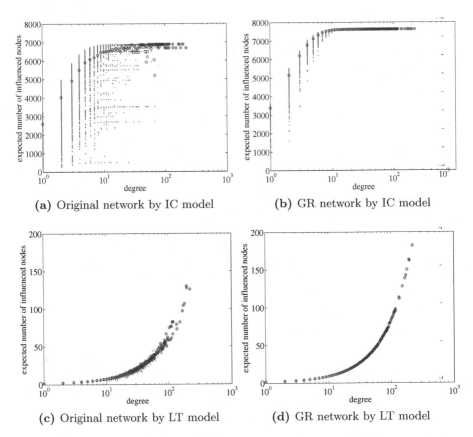

(a) Original network by IC model   **(b)** GR network by IC model

(c) Original network by LT model   **(d)** GR network by LT model

Fig. 1. Comparison of information diffusion processes using blog network

the LT model over the original Wikipedia network, and Figure 1d shows those of the LT model over the GR Wikipedia network. Here the red dots and blue circles respectively stand for the levels of the influence degree of individual nodes and their averages for the nodes with the same degree.

In view of the difference between the information diffusion models, we can clearly see that although nodes with larger degrees influenced many other nodes in both of the IC and LT models, their average curves exhibit opposite curvatures as shown in these results. In addition, we can observe that the influence degree of the individual nodes based on the IC model have quite large variances compared with those of the LT model.

In view of the difference between the original and GR networks, we can see that compared with the original networks, the levels of the influence degree were somewhat larger in the GR networks. We consider that this is because the averages of shortest path lengths became substantially larger than those of the GR networks, especially for the IC model. In the case of the LT model over the GR network (Figure 1d), we can observe that the influence degree was almost uniquely determined by the degree of each node. As the most remarkable point, in the case

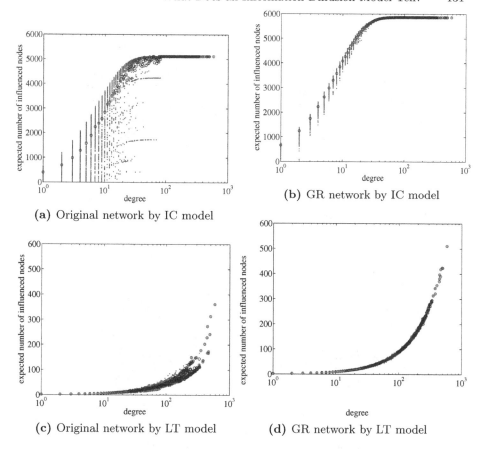

(a) Original network by IC model

(b) GR network by IC model

(c) Original network by LT model

(d) GR network by LT model

Fig. 2. Comparison of information diffusion processes using wikipedia network

of the IC model, we can observe a number of lateral lines composed of the individual influence degree over the original networks (Figure 1a), but these lines disappeared over the GR networks (Figure 1b).

4.5 Experimental Results Using Wikipedia Network

Figure 2 shows the same experimental results using the Wikipedia networks. From these results, we can derive arguments similar to those of the blog networks. Thus we consider that our arguments were substantially strengthen by these experiments.

We summarize the main points below. 1) Nodes with larger degrees influenced many other nodes, but their average curves of the IC and LT models exhibited opposite curvatures; 2) The levels of the influence degree over the GR networks were somewhat larger than those of the original networks in both of the IC and LT models; 3) The influence degree was almost uniquely determined by the degree of each node in the case of LT model using the GR network (Figure 2d);

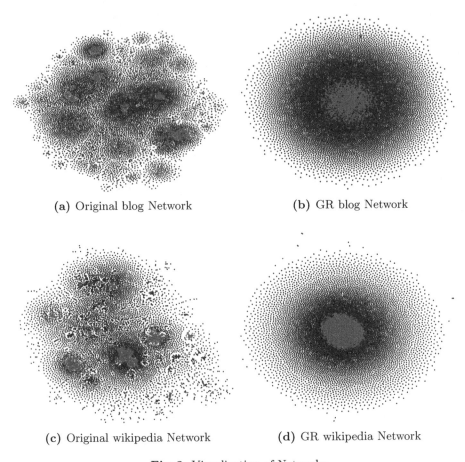

(a) Original blog Network   (b) GR blog Network

(c) Original wikipedia Network   (d) GR wikipedia Network

Fig. 3. Visualization of Networks

and 4) A number of lateral lines composed of the individual influence degree appeared in the case of IC model using the original network (Figure 2a).

4.6   Community Structure Analysis

Figure 3 shows our visualization results. Here, in the case of the blog networks, since the average degree was $\bar{d} = 6.6340$, we represented the nodes belonging to the 7-core subnetwork by red points, and others by blue points. Similarly, in the case of the Wikipedia networks, since the average degree was $\bar{d} = 25.8458$, we represented the nodes belonging to the 26-core subnetwork by red points, and others by blue points. These visualization results show that the nodes of higher core order are scattered here and there in the original networks (Figures 3a and 3c), while those nodes are concentrated near the center in the GR network (Figures 3b and 3d). This clearly indicates that the transformation to GR networks changes community structure from distributed to lumped ones.

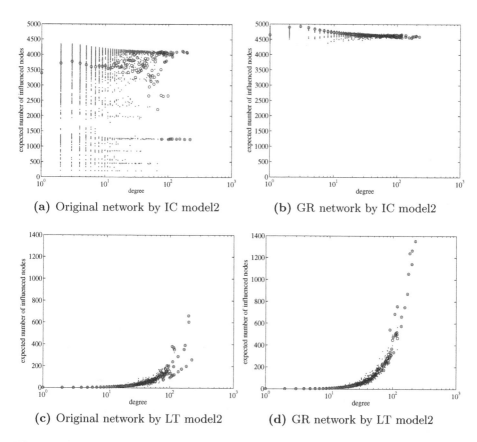

Fig. 4. Comparison of information diffusion processes using blog network by changed probability setting

Since the main difference between the original and GR networks are their community structure, we consider that a number of lateral lines appeared in the original networks using the IC model (Figures 1a and 2a), are closely related to distributed community structure of social networks. On the other hand, we cannot observe such remarkable characteristics for the LT model (Figures 1b and 2b). In consequence, we can say that community structure more strongly affects information diffusion processes of the IC model than those of the LT model.

4.7 Experimental Results by Changing the Parameter Settings

We consider to change the parameter settings to explore the intrinsic characteristics of the information diffusion models. First, we modify the IC model so as to roughly equalize the expected number of nodes influenced from any information source node. In the previous experiments, since we assigned the same diffusion probability to all links, the nodes with larger degrees have more advantage in

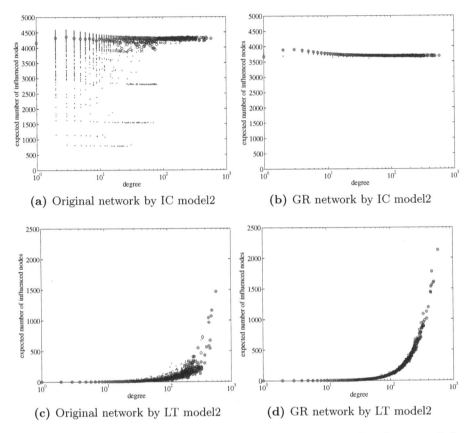

(a) Original network by IC model2   **(b)** GR network by IC model2

(c) Original network by LT model2   **(d)** GR network by LT model2

Fig. 5. Comparison of information diffusion processes using wikipedia network by changed weight setting

diffusing information than those with smaller degrees. In order to alleviate such a variation, we specified the diffusion probability for each directed link (u, v) as follows:

$$\beta_{u,v} = \frac{1}{|A_G(u)|}$$

Second, we modify the LT model so as to raise the expected number of nodes directly influenced from infomation source nodes with larger degree, just like the IC model using the common diffusion probability. To this end, we specified the weight for each directed link (u, v) as follows:

$$\omega_{u,v} = \frac{A_G(u)}{\sum_{x \in \Gamma(v)} |A_G(x)|}$$

Figures 4 and 5 show the experimental results, where we employed the same experimental settings used in the previous experiments, except for the above parameter settings. From these results, we can similarly confirm the fact that the community structure more strongly affects information diffusion process of the IC model than those of the LT model. We believe that this observation strengthens our claim. Here, as we expected, the average curves of the IC model show almost flat lines, and the number of influence nodes in the LT model becomes larger than previous experimental results.

5 Conclusion

In this paper, we proposed a new scheme for empirical study to explore the behavioral characteristics of representative information diffusion models such as the Independent Cascade model and the Linear Threshold model on large networks with different community structure. The proposed scheme consists of two parts, i.e., GR (generalized random) network construction from an originally observed network, and plotting of the influence degree of each node based on an information diffusion model. Using large real networks, we empirically found that our proposal scheme uncovers a number of new insights. Most importantly, we showed that community structure more strongly affects information diffusion processes of the IC model than those of the LT model. Our future work includes the analysis of relationships between community structure and information diffusion models by using a wide variety of social networks.

Acknowledgments

This work was partly supported by Asian Office of Aerospace Research and Development, Air Force Office of Scientific Research, U.S. Air Force Research Laboratory under Grant No. AOARD-08-4027, and JSPS Grant-in-Aid for Scientific Research (C) (No. 20500147).

References

1. Adar, E., Adamic, L.: Tracking information epidemics in blogspace. In: Proceedings of the 2005 IEEE/WIC/ACM International Conference on Web Intelligence, pp. 207–214 (2005)
2. Balthrop, J., Forrest, S., Newman, M.E.J., Williampson, M.W.: Technological networks and the spread of computer viruses. Science 304, 527–529 (2004)
3. Gruhl, D., Guha, R., Liben-Nowell, D., Tomkins, A.: Information diffusion through blogspace. In: Proceedings of the 13th International World Wide Web Conference, pp. 107–117 (2004)
4. Maximizing the spread of influence through a social network. In: Proceedings of the 9th ACM SIGKDD International Conference on Knowledge Discovery and Data Mining, pp. 137–146
5. Kimura, M., Saito, K., Nakano, R.: Extracting influential nodes for information diffusion on a social network. In: Proceedings of the 22nd AAAI Conference on Artificial Intelligence, pp. 1371–1376 (2007)

6. Kimura, M., Saito, K., Motoda, H.: Minimizing the spread of contamination by blocking links in a network. In: Proceedings of the 23nd AAAI Conference on Artificial Intelligence, pp. 1175–1180 (2008)
7. Newman, M.E.J.: The structure and function of complex networks. SIAM Review 45, 167–256 (2003)
8. Newman, M.E.J., Forrest, S., Balthrop, J.: Email networks and the spread of computer viruses. Physical Review E 66, 035101 (2002)
9. Newman, M.E.J., Park, J.: Why social networks are different from other types of networks. Physical Review E 68, 036122 (2003)
10. Seidman, S.B.: Network Structure and Minimum Degree. Social Networks 5, 269–287 (1983)
11. Yamada, T., Saito, K., Ueda, N.: Cross-entropy directed embedding of network data. In: Proceedings of the 20th International Conference on Machine Learning, pp. 832–839 (2003)

Web Mining for Malaysia's Political Social Networks Using Artificial Immune System

Ahmad Nadzri Muhammad Nasir, Ali Selamat, and Md. Hafiz Selamat

Intelligent Software System Research Laboratory, Faculty of Computer Science &
Information System, Universiti Teknologi Malaysia, 81310, Skudai, Johor, Malaysia
anmn85@gmail.com,
{aselamat,mhafiz}@utm.my

Abstract. Currently, politic is one of the critical and hot issues in Malaysia.
There are many aspects that are related with politics. These politic aspects are
mainly about spreading rumors and information through cyberspace that created
positive and negative effects to the political situations in Malaysia. There are
many branches of cyberspace that can be used by the Internet users to channel
their opinions such as websites, forums and blogs which contain the political issues in Malaysia. In this paper, we have analyzed an adaptive model of web
mining using Artificial Immune System (AIS) to retrieve the list of URLs that
have relevant information on political issues in Malaysia. In addition, we have
also used a concept of social network analysis in order to understand the relationships among the websites and web pages related to the political issues in
Malaysia. The results from this model have been used to analyze the social
networks and the impact of online community which contribute to the outcome
of Malaysia's 12th general election.

Keywords: Artificial Immune System, Social Network Analysis, Political Parties, General Elections.

1 Introduction

The ease use of many online social network applications on the Internet such as
community sites, e-business sites, knowledge sharing sites and blogs has come out
with the creation of online social networking services. With all of these applications,
many web users can create their own community website using web application such
as YouTube [15], Friendster [16], and etc. All of these websites can create a structure
of the social relationship between each member that will cluster a group of people.
With these relationships they can share many things such as their own profiles, vid-
eos, pictures and articles. The social network has created a medium to enable web
users to introduce new ideas to the other users and it helps in building the relationship
between individuals in community. Many researchers have performed analysis of
social network in many different areas such as online marketplaces [12], online community [7, 10, 11], etc.

Blog is a kind of web application that is used for publishing bloggers opinions
that are written by the blog owners or other bloggers. The content of the blog can be

D. Richards and B.-H. Kang (Eds.): PKAW 2008, LNAI 5465, pp. 137–146, 2009.

categorized and sorted in the reverse chronological order [9]. With the power that bloggers and authors of the website have in the cyberspace, they can make an impact to the political situation in some countries like Malaysia. The rumors and information that bloggers provide can give bad and good influences to the readers. Furthermore, this kind of information has become an important factor that influenced the results of Malaysia's 12th general election. As a result, the opposition party in Malaysia has won many chairs in the parliaments and managed to conquer five states [25, 26]. The Prime Minister of Malaysia, Mr. Abdullah who has accepted the fact said that bloggers is one of the reasons why the coalition governments did badly in Malaysia's 12th general election [8].

In this paper, we analyze the impact of online community towards the results of Malaysia's 12th general election using artificial immune system (AIS). The result from the AIS will then be used to form a social network structure of on-line communities in order to analyze the connection of relevant web pages based and user's opinions.

2   Problem Background

The result of Malaysia's 12th general election has given a substantial impact to the Malaysia history and political situation in Malaysia. From the result, the opposition parties in Malaysia that have been represented by the Parti Keadilan Rakyat (PKR), Democratic Action Party (DAP) and Parti Islam Se-Malaysia (PAS) has won many seats in the parliaments and five states have been controlled by the PKR [25, 26]. There are many reasons and speculations that have been made by the political analyst towards the election results. One of the reasons is the widely use of cyberspace by the opposition party leaders to reach the voters.

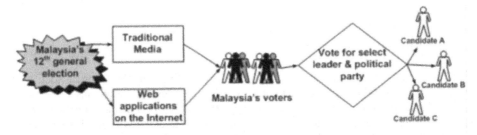

Fig. 1. Relationship between traditional media, web applications and Malaysia's voters towards Malaysia's 12th general election

The usage of Internet by the PKR which is the opposition party has created a huge impact to the result of Malaysia's 12th general election. There are some examples of popular blogs by the opposition party leaders such as Anwar Ibrahim [18] and Lim Kit Siang [20] which published their own stories or opinions widely through internet. This characteristics of web such as minimum cost, easy to maintain and offers anyone with an agenda a platform to preach have enable the PKR leaders to utilize it as a medium to reach the potential voters which is mainly the youngsters.

From the analysis that has been done by the Centre for Public Policy Studies (CPPS) [24], the results of the Malaysia 12th general election strongly has been affected by the traditional media and web applications. As for web applications on the Internet, people used forum, blogs and news portal to get the information. The internet medium has enabled the voters to watch, read and analyze all the stories and news about political issues in Malaysia via traditional media channel and web applications on the Internet. They can also read news and opinions that have been published by the bloggers which turn to be the candidates of the general elections.

Based on the analysis made by CPPS, the Malaysian opposition party such as DAP, PKR and PAS have used the web communities such as Youtube [15] and Friendster [16] as a tool to form a group in finding sources for their funds. They do believe that by using these web communities, they could develop a strong connection between them and widen their communities. Consequently, their strategy to influence the voters has affected the results of Malaysia's 12th general election. Fig. 1 shows us the relationship between Malaysia's 12th general election, traditional media, web applications and Malaysia's voters. This figure is based on our observation to the political issues in Malaysia. Based on this case study, our motivation is to find websites and blogs from other political leaders from both political parties that are related to the political issues in Malaysia. Then we want to analyze the connections between these websites and blogs and understand how it can make a big impact to the Malaysia's 12th general election and influence the voters on deciding their future government.

3 Related Works

In this paper, we referred to several papers to get the ideas on how to analyze the social network and the architecture of AIS that we applied in our web mining system. Artificial Immune System is defined by de Castro and Timmis [4] as the adaptive systems, inspired by the theoretical immunology and observed immune functions, principles and models, which are applied to problem solving. AIS used field of biological that inspired computing which can exploit theories, principles, and concept of modern technology. Some of the features of AIS are pattern of recognition, diversity, distributivity, self-organization and noise tolerance. There are many researchers that used AIS as a technique to solve problems like automatic timetabling [6], recommender for websites [2], etc.

The main idea of proposing the web mining comes from the Artificial Immune System for finding an interesting information discovery on the web (AISIID) [1]. AISIID is an application that has been created to find the relevant and interesting information for the user's search which also fulfils at least one of the criteria like novel, surprising and unexpected. From these web pages, user will find much more information that meets the subject or criteria that users want. AISIID uses a population of artificial immune cells and processes that been inspired by the clonal selection algorithm for both search and rank web pages. This AISIID use a WordNet [14] as a tool that help to expand the keywords and improve the performance of information retrieval.

All web pages that have been discovered need to be understood because all web pages are in the same subject and probably have a connection between them. From our study, we found that in order to understand the connection between all of the web pages, we need to build the social network of a group of these web pages. Social network can be described as a picture that shows the connection or social relationship in terms of nodes and ties. In social networks there are two aspects that played an important role in presenting the graph which are nodes and ties. Nodes are the individual in the network and ties are the relationship between the actors. In social network, node is exposing to various information if there are more connections frequency between nodes in the aspect of connection [3]. In the connection, nodes can be more influential and probably can be influenced by others. When the connection frequency is high, the information will be spread more quickly. Therefore, social network can give a big impact to the international culture and global economy if the connection and its structure are strong [13].

4   Web Mining Using Artificial Immune System

This section will explained more on the process of how the web mining system can be used with Artificial Immune System (AIS) technique. This web mining is different from the AISIID and other famous search engines such as Yahoo [22] and Google [23]. In AISIID, it uses several seed URLs from the users and ranks the words by word frequency to generate keywords. Different with Google and Yahoo, these two search engines take an amount of keywords from users and give the set of results in the shortest time. The results of Google and Yahoo will be ordered by the rank of popularity and this result will returned to the user typically in thousands of documents or more. Sometimes, it is difficult for the user to search for documents that are relevant to the user's subject. Thus, we believed that AISIID can help to solve the problem that might faced by the users when they have a lot of documents to be reviewed.

The idea of proposing the web mining using AIS method came out after we realized that there are several disadvantages of AISIID. In this paper, we believed that AISIID is not the best technique to find relevant website and web pages about political issues in Malaysia. The problem arises when the user wants to find subject in Malay words, because WordNet cannot recognize and expand the generated keywords by AISIID. This problem makes the result from AISIID is not really relevant to the user. AISIID also used text surrounding to select the hyperlinks in every page so that the artificial immune cell allows the search space to be explored and to aid search diversity. To select the best hyperlinks, all of the hyperlinks in the web pages are weighted using a measure of relevance text surrounding and if the texts which surround the hyperlinks are relevant, the hyperlinks will be selected. The problem occurs when some of the texts which surround the hyperlink may not be relevant but the content of the hyperlink may be relevant to the user's subject. All of these problems can be solved by using our model of web mining with the AIS method.

4.1   Overview of Web Mining Using Artificial Immune System

Our web mining with AIS uses a population of cells and processes inspired by the clonal selection model for searching and ranking web pages. Clonal selection model is the model that gives the best performance on learning and maintenance of high quality memory [3]. Our web mining has enhanced several aspects such as uses the users'

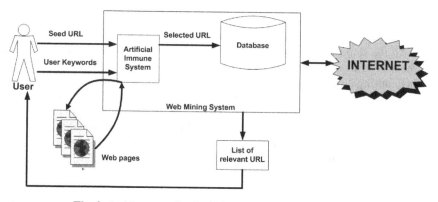

Fig. 2. Architecture of web mining system using AIS method

keywords from user to search web pages that contain relevant information in Malay language that fulfils the user's subject. Automated feedback mechanism is applied in this web mining system and use affinity measure to evaluate all appear hyperlinks in web pages to maximize search. We believed that these innovative aspects that have been applied in our web mining using AIS can help to solve the problems that we described in the Section 2. Fig. 2 shows the architecture of the web mining system by using AIS.

The aim of our web mining system by using AIS is to give a small group of web pages that are highly relevant to the user's subject. This web mining system needs a kind of inputs which are seed URLs and user keywords to get the results those summaries the user's subject to get the results. This seed URLs and user keywords must be relevant to the user subject so the web mining can returned back the relevant web pages. From the starting page based on the seed URLs, this web mining will crawl all the links in the seed URL page. Each web page that has been discovered based on URL is recognized and this web mining will select the web pages that contain relevant information to the user's subject. This selection is based on calculation of affinity that helps the system to classify the population of relevant URL. This web mining system will stop when the criterion is met and after that the system will list the entire relevant URL which been ordered by the affinity. The result can be visualized in form of social network. The visualization helps to show the user who is the key player in the social network, pattern of links and relationship between web pages related to political issues in Malaysia.

4.2 Algorithm Description

This section will explain about the process of web mining which is based on the AIS framework of [4] i.e., representation, affinity measure and process.

Representation. In this section, we will explain about the artificial immune cell job as an agent in this web mining. The main task of artificial immune cells is to represent relevant information in the result of web mining process. Artificial immune cell will present information to the user in three types.

First is a summary of relevant information for every web page which is called antigens that will be detected by the cell to determine whether it is relevant to the user's subject. All the summary of relevant information for every web page will be saved in a document as strings. It assumed that the result has met the user's knowledge and will not change during a single run of the system. To determine whether the web pages are relevant, every cell must carry a set of words relevant (user keywords) to the user's subject. Second, a location of web pages on the web (a list of URL). If current web page is relevant, the cell will hold the URL and the URL will be saved into the database. The third is a count of affinity measure where cell will count all the words in the current web page that is similar to the user keywords.

$$affinity = \frac{\sum_{i=1}^{|UK|} \delta i}{|UK|} \quad where \quad \delta i = \begin{cases} 1 \ if \ UK_i \in W \\ \hline 0 \ otherwise \end{cases} \tag{1}$$

Affinity function. Basically, the result from the web mining system using AIS method uses a part of the mathematical formulation from AISIID where the formulation will be adjusted to count the relevance words that are similar to the keywords that have been given by user. In Eq. 1, the number of words in the user keywords (UK) that also appear in webpage are counted and divided by the length of the user keywords (UK). This formulation is used to access the affinity between seed page and other pages. The result from the formulation will be used as a measurement for every web page.

Process. The process of web mining system start when user enters the seed URLs and keywords. Then the web mining system automatically initializes the population and generates the cells to be placed at the starting page. This system then used user keyword as active cells (B-cells) and presents all hyperlinks in the starting web page. In the theory of immune system, there are cells called B-cells which help to recognize whether the antigens that enter into the human body are safe and B-cells will remove all pathogens like dangerous viruses, and bacteria that will invade a host [4].

In the process of initializing the population, the content of seed URL must go through another process which is called preprocessing. This process strips all HTML tags and grabbed all the hyperlinks. All the hyperlinks then will be saved into the database with their reference and content. In a single run of the system, the AIS used all the active immune cells to help rank all the words in the seed page by word frequency. In this case, if the word "Anwar" is most frequently appears in the web page than other keyword, so this method will recognize "Anwar" word as the first rank.

After that, all the saved hyperlinks are evaluated based on the contents that must through some processes such as preprocessing and word frequency. The evaluation of every URL's content is using the active cells and mathematical function in Eq. 1 to access the affinity between current URL and seed URL. Then, every URL is categorized in to two groups. The high affinity group is the group that the content of web page is similar to the seed page while the low affinity is not similar to the seed page. To select the relevant URLs, the content of URL must be relevant to the user's subject. URL that has the high affinity is selected to be placed in the best population and all of the artificial cells in the web pages are stimulated. Stimulation above a threshold

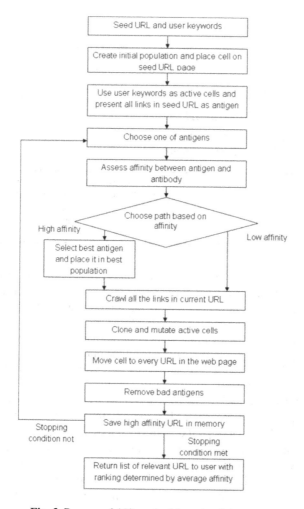

Fig. 3. Process of AIS method in web mining system

causes the cell to clone and mutate. Cells will go to another URL by the following hyperlinks that appear in the high affinity URL. For low affinity URL, the AIS will only grab all the hyperlinks. This part of cloning and mutation of active cells help for maximizing the search of relevant web pages. After that, all the clones of active cells will freely go to every hyperlink and do evaluation. The process of AIS stops when all the hyperlinks are evaluated by active cells, then all the hyperlinks will be saved into the database and will return it to the user. Fig. 3 shows the process of AIS in the web mining system.

4.3 Experimental Results and Analysis

The aim of this experiment is to get a small number of highly relevant web pages about political issues in Malaysia by using our web mining with AIS. The results from

Table 1. Result from web mining using AIS

Seed URL	Web pages discovered	Relevant Web page by web mining using AIS	Percent of relevant (%)
http://www.mohdalirustam.com	873	400	45.82
http://www.anwaribrahimblog.com	947	400	42.24

Table 2. Number of website/blog that connect with seed URL

Seed URL	Number of website/blog that connected to seed URL	Number of relevant website/blog to political issue in Malaysia
http://www.mohdalirustam.com	56	49
http://www.anwaribrahimblog.com	74	67

the experiment will be used to analyze the social network between the communities of governance party and the opposition party. In this experiment, we need URL website or blog from the top leader in every party as the representative. We believed that by choosing the URL website or blog from the top leader in the party, it will show us the connection of website or blog by other leaders in the party and the web pages about the political issues. So we choose two URLs from the opposition party leader which is Datuk Seri Anwar Ibrahim's blog [18] and Datuk Seri Mohd Ali Mohd Rustam's blog [19] from the governance party.

These two selected URL will be used as seed URLs and the user keywords in Malay words are "rakyat", "politik" and "Malaysia". In this experiment, we limit to 400 relevant web pages that related to the political issues in Malaysia. The reason for limiting the relevant web pages is we want to give a small group of web pages to the user. Table 1 shows the result from web mining with AIS. Table 2 gives the number of websites or blogs that have connections with the seed URLs and the number of relevant websites or blogs to the political issues in Malaysia that have connections with the seed URLs.

After a single run, there are 873 web pages that have been discovered to get 400 web pages which are related to the political issues in Malaysia for Datuk Seri Mohd Ali's blog. This blog is connected with 56 URLs of website or blog but only 49 URLs that have been judged by us which are relevant to the political issues in Malaysia. As for Datuk Seri Anwar's blog, there are 947 web pages been discovered to get 400 web pages that are relevant to the political issues in Malaysia and there are 67 URLs of website or blog from 74 URLs that have been judged by us are relevant to the political issues in Malaysia. The result also leads us to other websites or blogs that contain information about political issues in Malaysia.

In Figs. 5 and 6, it shows the social network of http://www.mohdalirustam.com and http://www.anwaribrahimblog.com. From the figures, it shows that the Datuk Seri Mohd Ali's blog do not have a strong community in cyberspace compared to Datuk Seri Anwar's blog. These two blogs also are the key to other websites or blogs by the

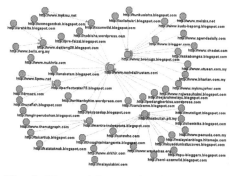 

Fig. 5. The social network of http://www. mohdalirustam .com

Fig. 6. The social network of http://www. anwaribrahimblog.com

other leaders in every party. From the analysis, we believed that by having a strong connection in the online community, it can give a big impact to the global situation. It can be proved with the strong online community by leaders in the opposition party which influenced the voters to vote them and won many chairs in the parliaments.

5   Conclusion

From the result, we believed that arises of web applications like blog and online portal that provide the information and news about political issues gave a big impact to the results of Malaysia's 12th general election. The web mining using AIS technique helps the researcher to analyze the relationship between blogs that contains relevant information to political issues in Malaysia. In the future, we planned to enhance our web mining system with AIS to do more better searching for relevant information. For social network, we want to analyze more on social network in term of impact to the political situation in Malaysia. Visualization technique can also be added to show the connection between blogs for user understanding.

Acknowledgments. The authors wish to thank the reviewers for their helpful suggestions and members of the Intelligence Software Systems Lab (ISSLab) at Faculty of Computer Science & Information Systems, Universiti Teknologi Malaysia (UTM). This work is supported by the Ministry of Science and Technology and Innovation (MOSTI), Malaysia and Research Management Centre, Universiti Teknologi Malaysia (UTM) under the Vot 79267.

References

1. Secker, A., Freitas, A.A., Timmis, J.: AISIID: An artificial immune system for interesting information discovery on the web. Applied Soft Computing 8, 885–905 (2008)
2. Morrison, T., Aickelin, U.: An Artificial Immune System as a Recommender for Web Sites. In: Proceedings of the 1st Internal Conference on Artificial Immune Systems (ICARIS 2002), Canterbury, UK, pp. 161–169 (2002)

3. de Castro, L.N., Von Zuben, F.J.: The Clonal Selection Algorithm with Engineering Applications. In: Proceedings of GECCO 2000, Las Vegas, USA, pp. 36–37 (2000)
4. Castro, L.N., Timmis, J.: Artificial Immune Systems: A New Computational Intelligence Approach. Springer, Heidelberg (2002)
5. Hunt, J.E., Cooke, D.E.: Learning using an artificial immune system. Journal of Network and Computer Applications 19, 189–212 (1996)
6. He, Y., Hui, S.C., Lai, E.M.-K.: Automatic Timetabling Using Artificial Immune System. In: Megiddo, N., Xu, Y., Zhu, B. (eds.) AAIM 2005. LNCS, vol. 3521, pp. 55–65. Springer, Heidelberg (2005)
7. Warmbrodt, J., Sheng, H., Hall, R.: Social Network Analysis of Video Bloggers' Community. In: Proceedings of the 41st Hawaii International Conference on System Science (2008)
8. Blogging the in-thing for politicians,
   http://www.nst.com.my/Current_News/Saturday/National/
   2217630/Article
9. Wright, J.: Blog Marketing. McGraw-Hill, New York (2006)
10. Yao-Jen, C., Yao-Sheng, C., Shu-Yu, H., Chiu-Hui, C.: Social Network Analysis to Blog-based Online Community. In: International Conference on Convergence Information Technology (2007)
11. Li, F.-r., Chen, C.-h.: Developing and Evaluating the Social Network Analysis System for Virtual Teams in Cyber Communities. In: Proceedings of the 37th Hawaii International Conference on System Science (2004)
12. Kumar, P., Zhang, K.: Social Network Analysis of Online Marketplaces. In: IEEE International Conference on e-Business Engineering (2007)
13. Gatson, M.E., desJardins, M.: Social Network Structures and their Impact on Multi-Agent Dynamics. American Association for Artificial Intelligence (2005)
14. WordNet, http://wordnet.princeton.edu
15. YouTube, http://www.youtube.com
16. Friendster, http://www.friendster.com
17. Internet World Stats, http://www.internetworldstats.com/stats3.htm
18. Ibrahim, A.: http://www.anwaribrahimblog.com
19. Kacamata, D., Rustam, M.A.: http://www.mohdalirustam.com
20. Siang, L.K.: http://blog.limkitsiang.com
21. Malaysiakini, http://www.malaysiakini.com
22. Yahoo, http://www.yahoo.com
23. Google, http://www.google.com
24. Centre for Public Policy Studies (CPPS), http://www.cpps.org.my
25. Election Commission of Malaysia, http://www.spr.gov.my
26. The Star Online - General Election (2008), http://thestar.com.my/election

Accessing Touristic Knowledge Bases through a Natural Language Interface

Juana María Ruiz-Martínez, Dagoberto Castellanos-Nieves,
Rafael Valencia-García, Jesualdo Tomás Fernández-Breis,
Francisco García-Sánchez, Pedro José Vivancos-Vicente,
Juan Salvador Castejón-Garrido, Juan Bosco Camón,
and Rodrigo Martínez-Béjar

Facultad de Informática, Univeridad de Murcia, Spain
Universidad de Murcia 30071 Espinardo (Murcia). Spain
{jmruymar,dcastellanos,valencia,jfernand,frgarcia,rodrigo}@um.es
VOCALI SISTEMAS INTELIGENTES S. L.
Edificio CEEIM. Campus de Espinardo. 31000 Espinardo (Murcia). Spain
pedro.vivancos@vocali.net, juans.catejon@vocali.net

Abstract. In order to realise the Semantic Web vision, more and more information is being made available in formal knowledge representation languages such as OWL. Unfortunately, the gap between human users who want to retrieve information and the Semantic Web remains unsolved. This paper presents a method for querying ontological knowledge bases from natural language sentences. This approach is based on the combination of three key elements: Natural Language Processing techniques for analyzing text fragments, ontologies for knowledge representation and Semantic Web technologies for querying ontological knowledge bases. The results of the application of this approach in the e-tourism domain are also reported in this paper.

1 Introduction

Motivated by the new advances and trends in Information Technologies (IT), an increasing number of tourism industrial operators offer their products and services to their customers through online Web services. Similarly, regional and local administrations publish tourism-related information (e.g. places of interest, hotels and restaurants, festivals, etc.) in world-accessible websites. Hence, the tourism industry is becoming information-intensive, and both the market and information sources heterogeneity are generating several problems to users, because finding the right information is becoming rather difficult. Software applications might support users in their searches, but most of the available information is represented in natural language, and this is difficult to understand and exchange by software programs.

Hence, mechanisms for supporting users in their information retrieval activities are needed. The Semantic Web enables better machine information processing, by structuring web documents, such as tourism-related information, in such

D. Richards and B.-H. Kang (Eds.): PKAW 2008, LNAI 5465, pp. 147–160, 2009.

a way that they become understandable by machines [12]. These technologies might help us to provide the intended support to users. In the Semantic Web, knowledge is represented by means of ontologies. An ontology is viewed in this work as a formal specification of a domain knowledge conceptualization [10]. In this sense, ontologies provide for a formal, structured knowledge representation schema that is reusable and shareable.

Concerning semantic information retrieval, currents approaches for querying ontologies use formal ontology-based queries, such as SPARQL[23], RDQL[20], RQL[22] and OWLQL[13]. These languages return tuples of ontology values that satisfy the queries. However, these formal queries are too complex to be used for non-expert users. Consequently, developers usually implement these queries in forms.

Thus, the successful application of these technologies in the eTourism domain is based on the availability of tourism ontologies, which would provide a standardized vocabulary and a semantic context. In the last few years, several ontologies for tourism have been developed. Particularly, in our research we have combined the Protégé travel ontology [14] and OnTour [19]. The resulting ontology was extended by adding new classes mainly related to restaurants.

In this work, a semantic processing-based approach for retrieving information from natural language queries is presented. For this, the textual fragments that constitute the query are pre-processed by means of Natural Language Processing techniques, including POS-tagging, lemmatizing and search of synonyms. In this approach, Classes, Object Properties, Datatype Properties and Individuals can be annotated by using the available annotation labels. In this way, the text content can be briefly described, so allowing for querying OWL repositories by using natural language.

1.1   e-Tourism Ontology

A major bottleneck in ontology building is that of knowledge acquisition. The development of an ontology for a particular domain is a time-consuming task while a shared vocabulary is necessary for obtaining an agreed ontology.

Several ontologies for e-tourism have been developed. The European project called Hi-Touch used the ontology created by the Mondeca working group. This ontology contains six categories of classes: activity, certification, environment, ethics, logistics and philosophy [6]. It also includes tourism concepts from the Thesaurus on Tourism and Leisure Activities developed by the World Tourism Organisation (WTO), which is a multilingual thesaurus in English, French and Spanish that provides a standard terminology for tourism [25].

DERI's e-Tourism Working group has created a tourism ontology named On-Tour, whose concepts have also been extracted from WTO's thesaurus. This OWL ontology describes the main conventional concepts for tourism such as Accommodation or Activity, together with other supplementary concepts like GPS-coordinates or Postal Address [19].

The Harmonise European project has developed the Interoperable Minimum Harmonisation Ontology (IMHO). This is formed by a limited number of the most representative concepts of the tourism industry, so allowing for information exchange between tourism actors [17].

The SEmantic E-tourism Dynamic (SEED) packaging research laboratory has developed the Ontology for Tourist Information Systems (OTIS), which is applicable exclusively to e-Tourism. SEED has been used for creating a "Dynamic Packaging" [5], which is a custom package for consumers, created automatically by using Internet technologies.

Another e-Tourism ontology is the Australian Sustainable Tourism Ontology (AuSTO), which has been developed at the University of Melbourne and provides a mechanism for planning trips by matching the user requirements with the vendor offers [12].

Finally, LA_DMS[11] is a comprehensive ontology for tourism destinations that was deployed for the Destination Management System (DMS). This system adapts information requests about touristic destinations to users' needs.

By considering the shortcomings of developing a new ontology from scratch, we have reused the ontology for e-tourism developed by Protg [14], adding new Restaurant-related classes and other properties from the OnTour [19] ontology. As a result, we have obtained an ontology which contains all the touristic information that will be used by the system (see fig. 1). The ontology has been implemented in OWL and its knowledge is accessed through the Jena API [1].

The rest of the paper is organized as follows. In Section 2, the natural language query process is described. The application of the process is illustrated through the example presented in Section 3. In Section 4, some results of this approach are shown. Finally, some conclusions are put forward in Section 5.

2   Overview of the Natural Language Query Interface

The framework used in this methodology is comprised of three main modules or phases (fig. 2): sentence preparation, knowledge entities search, and information search. Next, these modules are explained in detail. In addition to this the processes that take place in each phase are described.

2.1   Sentence Preparation Phase

In this phase, the query written in natural language is pre-processed. For this, three main steps are carried out. First, a POS-Tagging process is performed. This allows the system to identify the grammar category of each word in the sentence and removes the semantically-meaningless words. Then, the system identifies the stem of each word by means of a lemmatizing process. Finally, the synonyms related to the tourism domain are obtained. GATE tools [18] are employed for this pre-processing phase.

2.1.1   POS-Tagging

The main objective of the first stage of this phase is to obtain the grammatical category of each word in the current sentence. The existence of semantically

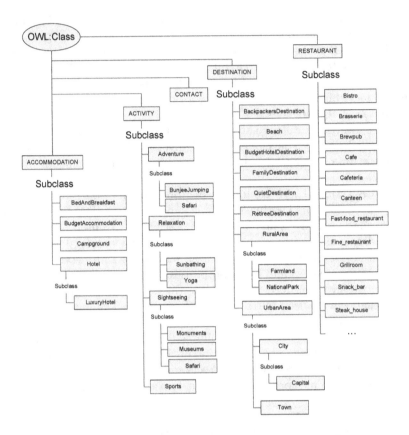

Fig. 1. Part of the class design in the e-tourism ontology

meaningless words is assumed, and these usually belong to the following grammar categories: Preposition, Conjunction, Interjection, Particle, Pronoun and Determiner. Words belonging to these types are removed from the query in this phase.

Let us suppose the sentence "*I [PRP] want [VBP] to [TO] visit [VB] the [DT] most [RBS] important [JJ] Paris [NNP] tourist [NN] attractions [NNS].*"

After this step, the following partial result would be obtained:

want [VBP] visit [VB] most [RBS] important [JJ] Paris [NNP] tourist [NN] attractions [NNS].

As it can be observed, the semantically meaningless words have been removed from the query.

2.1.2 Lemmatizing

In this work, "lemma" refers to a headword or heading in any kind of dictionary, encyclopedia, or commentary. Therefore, in this phase the system obtains the lemma of each query word in this phase.

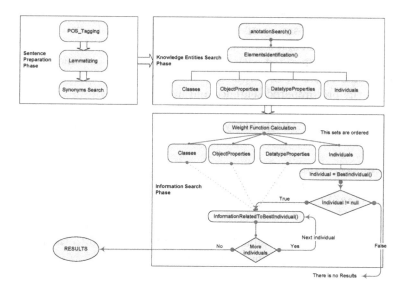

Fig. 2. Natural language query interface

The previous example, only the word "attractions" must be transformed into "attraction".

Some languages, such as Spanish, are highly inflectional, so the lemmatizing process is very useful, because it enables to identify gender and grammatical number, irregular verbs or enclitic pronouns. In this way, language canonical forms can be obtained.

2.1.3   Synonyms Search
In this phase, the system searches for possible synonyms for each query word using a tourism thesaurus.

The ontology has been enriched with synonyms of the terms assigned to the classes of ontology by using the Multilingual Thesaurus on Tourism and Leisure Activities [25]. In particular, the thesaurus descriptors and non-descriptors terms have been included as labels of these classes.

2.2   Knowledge Entities Search Phase

The objective of this phase is the identification of the knowledge entities that appear in the query and their classification into the four ontological categories included in our approach: Classes, DataTypeProperties, ObjectProperties and Instances.

This phase is composed of two main functions, namely, annotationSearch() and elementsIdentification(), which are described next.

2.2.1   Annotation Search

The "rdfs:label" is an instance of "rdf:property" that may be used to provide a human-readable version of a resource name. In this work, all the resources in the ontology have been annotated with the label descriptor. This function searches in the label annotations of the ontology for all the elements that contain any of the words in the pre-prepared sentence and in their synonyms.

This function works as follows. First, each label annotation is lemmatized. Then, the following SPARQL query is executed for each word and for each synonym detected in the previous phase:

```
SELECT ?entity, ?label WHERE{
?entity rdfs:label ?label FILTER
regex(str(?label),current)}
```

where *current* represents the words or synonyms obtained until this phase. At the end of this phase, the possible knowledge entities related to the meaningful query words and their synonyms are obtained.

2.2.2   Elements Identification

Once the annotationSearch function has been applied, the system gets a set of knowledge entities and their corresponding labels, all being related to the query. After that, the *elementsIdentification* function classifies all these knowledge entities into four different sets:

- Classes: Knowledge entities identified in the query that are classes in the ontology.
- ObjectProperties: Knowledge entities identified in the query that are object-Properties in the ontology.
- DataTypeProperties: Knowledge entities identified in the query that are dataTypeProperties in the ontology.
- Individuals: Knowledge entities identified in the query that are individuals in the ontology.

2.3   Information Search Phase

In this phase, each knowledge entity is assigned a score through a Weight function and the Individual with the best score is obtained. After that, the possible instances the user is asking for in the query are obtained by using the method implemented in the InformationRelatedToBestIndividual function.

2.3.1   Weight Function

Once the knowledge entities potentially related to the query have been identified and classified, the system has to select the best answer. The sets obtained in the previous phase are sorted by applying this function, which also gets an estimated weight for each knowledge entity according to the query.

The weight function works as follows. First, the system obtains the label annotations for each knowledge entity with a hit in the annotationSearch. Then, for each knowledge entity, the following formula is applied to calculate its respective score:

$$score_i = \frac{\displaystyle\sum_{j=0}^{j=number\_of\_annotations} \dfrac{n\_equal\_tokens(annotation_j, query\_tokens)}{n\_tokens(annotation_j)}}{number\_of\_annotations} \tag{1}$$

For each annotation, the number of equal words is divided by the total number of tokens in the annotation. Next, all these scores are added and then, divided by the total number of annotations of the knowledge entity under question.

Others textual similarity functions between the lemmatized query and the annotations, like cosine similarity [21], could be applied.

2.3.2   Information Related to Best Individual

Once all the sets of knowledge entities have been ordered, the system gets the first (i.e. the best) individual in the list (using the BestIndividual function). Then, the information related to the individual asked in the query is obtained by using the InformationRelatedToBestIndividual function.

This function identifies the knowledge entities (Classes, ObjectProperties, DatatypeProperties and Individuals) that have been identified and that are related to this individual. At this point, there are four possible scenarios:

- Case 1: If the best individual is related to another individual whose class has been identified, then the individual is one of the knowledge entities the user asked for.
- Case 2: If the best individual has an ObjectProperty in the list, then the individuals related to the best individual through the referred ObjectProperty are obtained.
- Case 3: If the best individual has a DatatypeProperty in the list, then the DatatypeProperty and the value are shown.
- Case 4: If the best individual is related to another individual identified in the previous phase, then the best individual will only be recovered if this and the search expression have some tokens in common.

The system searches sequentially through the labels for all possible sentence combinations until a hit is made. The Search Phase is over when a result for the query is reached.

2.3.3   Search Result Display

The visualization of the results will vary depending on which knowledge entity is related to the retrieved information. If the information is only an individual, then the data stored in its rdfs:comment are displayed, obtaining a brief description of the individual.

If the best individual is related to an ObjectProperty and/or a Datatype-Property that has been identified by the weight function, then their respective rdfs:comment data will be also shown.

3   Example of Query Processing

In this section, how the proposed methodology works is illustrated through an example. Let us suppose that a user is searching for information about places to visit in Paris and formulates the following natural language query: "I want to visit the most important Paris tourist attractions".

First, the POS-tagger assigns grammatical categories to each word in the sentence in the following manner:

I [PRP] want [VBP] visit [VB] most [RBS] important [JJ] Paris [NNP] tourist [NN] attractions [NNS].

Now, the semantically-meaningless words such as prepositions, conjunctions, interjections, particles, pronouns and determiners, are removed. The final result of this phase is "Want visit most important Paris tourist attractions".

Next, the lemmatizing process is carried out and the inflected terms are reduced to their lemma. In our example, all the words remain unchanged, except for "Attractions", which is reduced to its singular form, that is, "Attraction".

After that, travel synonyms are searched for and extracted. In our example, the synonyms showed in Table 1 are obtained. Then, the query would be transformed into the following sentence:

"Want {visit, see, travel to} most important {Paris, City of Light, French capital, capital of France} {tourist attraction, places of interest, tourist places, sightseeing}"

Table 1. Synonums obtained

Word	Synonyms
Visit	See, Travel to
Paris	City of Light, French capital, capital of france
Tourist attraction	Places of interest, sightseeing

All the previous terms are used to search for the possible knowledge entities referenced in the travel ontology. Let us suppose that the knowledge entities in the ontology related to the query are those shown in fig 3. The ontology has an ObjectProperty "hasActivity" between the classes "DESTINATION" and "ACTIVITY". Besides, the individual *Paris* is related to the individuals *Louvre_Museum, Eiffel_Tower, Notre_Dame_Cathedral, Arch_of_Triumph* through this object property. Each knowledge entity has a set of rdfs:label annotations in two languages, Spanish and English, what permits to query the ontology in two different languages.

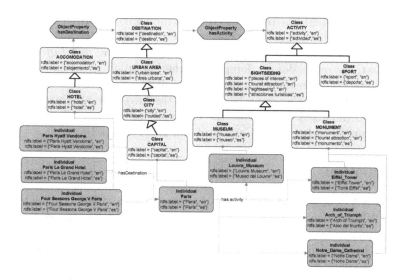

Fig. 3. Knowledge entities identified in our example

The terms in the modified query are then searched for in the rdfs:label annotations (knowledge entities search phase), obtaining the knowledge entities shown in Table 2.

Once the potential knowledge entities have been identified, the information search phase is executed. First, the weight function is applied to each identified knowledge entity shown in Table 2.

1. Linguistic expressions matching the user query are searched into the ontology annotation labels.
2. The number of words that are not present into the recovered label annotations is counted (unrecovered words).

In this example, the best individual is Paris. Then, other individuals connected to Paris, whose classes are also present in the weight function, are searched for.

Table 2. Results and knowledge entities which are linked to the retrieved labels

Knowledge entity	Label annotations	Match	No-match	Score
Class SIGHTSEEING	Places of interest	3	0	1
	Tourist attraction	3	0	
	Sightseeing	1	0	
Class MONUMENTS	Monuments	0	1	0,5
	Tourist attraction	2	0	
Individual of CAPITAL	Paris	1	0	1
Individual of HOTEL	Paris Hyatt Vendome	1	2	0,33
Individual of HOTEL	Paris Le Grand Hotel	1	3	0,25
Individual os HOTEL	Four Seasons George V Paris	1	4	0,20

As the best individual has relations to some individuals of the class "SIGHT-SEEING", these individuals are identified as a part of the result of the query (*Louvre_Museum, Arch_of_Triumph, Notre_Dame_Cathedral and Eiffel_Tower*).

After that, the methodology determines whether an object or datatype property related to the individual has been identified as a possible knowledge entity in the second phase. In the example, no properties have been obtained in the second phase, so the query only shows the following individuals: Louvre Museum, Arch of Triumph, Notre Dame Cathedral and Eiffel Tower.

4 Evaluation

The methodology described in this work has been employed for querying tourism-related OWL documents. Four PhD students of Computer Science were asked to design 20 queries that should be answerable by the information contained in the tourism web portal of the Region of Murcia (http://www.murciaturistica.es). The students were asked to first express the queries in natural language, and then to transform them into SPARQL. Both sets of queries were then executed and their results were compared. Table 3 shows a representative set of the queries designed in the experiment for both approaches.

Table 3. Some of the queries

NL query	SPARQL Query	Result
Tourist attractions in *Murcia*	SELECT ?sigthseeing WHERE { :Murcia has_activity ?sigthseeing ?sightseeing a :SIGHTSEEING}	Individuals of SIGHTSEEING which are related to the individual Murcia through the ObjectProperty "has_activity"
Hotels in Murcia with *golf course*	SELECT ?hotel WHERE { ?hotel has_destination :Murcia. ?hotel has_sevice :golf_course. ?hotel a :HOTEL}	Individuals of HOTEL which are related to the individual *Murcia* through the ObjectProperty "has_destination". The individuals of HOTEL are also related to the individual *golf_course* through the ObjectProperty "has_service"
Hotels near the *Cathedral*	SELECT ?hotel WHERE { ?city has_activity :*Cathedral*. ?hotel has_destination ?city. ?hotel a :HOTEL}	Individuals of HOTEL which are related to the individuals of CITY through the ObjectProperty "has_destination" and individuals of CITY related to the individual *Cathedral* through the ObjectProperty "has_activity"
Eating places in *Murcia*	SELECT ?restaurant WHERE { ?restaurant has_destination :*Murcia*. ?restaurant a :RESTAURANT}	Individuals of RESTAURANT which are related to the individual *Murcia* through the ObjectProperty "has_destination"

Table 4. Results for the queries shown in Table 3

Precision				Recall			
Student 1	Student 2	Student 3	Student 4	Student 1	Student 2	Student 3	Student 4
0.87	0.82	0.86	0.88	0.9	0.88	0.89	0.95

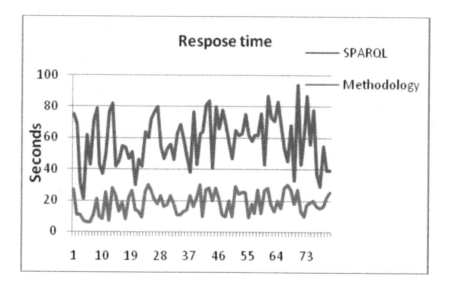

Fig. 4. Comparison between the response times using SPARQL and the Natural Language Interface

The evaluation of the results has been done through the precision and recall score. The precision score is obtained by dividing the amount of knowledge entities obtained by the natural language interface that have been obtained also by the SPARQL query, and the total amount of knowledge entities obtained by the natural language query. The recall score is obtained by dividing the amount of knowledge entities obtained by the natural language interface that have been obtained also by the SPARQL query, and the total amount of knowledge entities obtained by the SPARQL query.

precision = (correct knowledge entities obtained in NL query)/(total knowledge entities obtained in NL query)

recall = (correct knowledge entities obtained in NL query)/(total knowledge entities in SPARQL query)

The results obtained, shown in table 4, seem promising. In particular, the system obtained the best scores for student 4, with a precision of 0.88 and a recall of 0.95. In general the system obtains 90% for recall, but the system obtains some incorrect results (precision score). In Figure 4, we show the time spent for the 20 different queries

According to the results shown in Table 4, the quality of both approaches is similar, but the NLP approach has a lower response time than using SPARQL.

5   Discussion and Conclusion

In this paper, a method for querying OWL content in natural language has been presented. This approach is based on the analysis of the annotations of the ontology elements. The framework has three main phases: Sentence Preparation, Knowledge Entities Search and Information Search.

Other authors, like Bernstein [4] have dealt with querying OWL content in natural language. They provide a guided, quasi-natural language interface for accessing OWL knowledge bases that shows the possible vocabulary to be used to the user. This vocabulary is obtained from the vocabulary defined by the ontologies being currently under consideration.

In the Ontoligent Interactive Query (OntoIQ) project [2], natural language queries for evaluating domain ontologies are addressed. These queries are mapped onto predefined nRQL queries for using the RACER reasoner. Its main drawback is the limited linguistic expressiveness of the queries. Other approaches such as [8] transform natural language queries to predefined SPARQL queries in a similar way.

Current approaches for querying ontologies such as SPARQL, RDQL and OWLQL are too complex to be used for non-expert users. So, developers usually implement these queries in forms.

In [7], an interface that allows for querying tourism information from relational databases in natural language is presented. This interface translates user's query concepts into parameterized SQL fragments. An ontology is used to specify the concepts that are relevant in the application domain and to describe linguistic relationships like synonymy. Besides, the knowledge base contains the description of the parameterized SQL fragments that are used to build a single SQL statement representing the natural language query. This implies that, for each new database, a new knowledge base is necessary.

In our approach, the ontology contains not only basic concepts, but also specific concepts, attributes, relationships and instances. Moreover, all the information available for user queries is represented in the ontology. It is also possible to include subjective information related to knowledge entities in the ontology, what is an advantage with respect to others models. The tourism information available in the web is very subjective, because it includes a lot of opinions and recommendations. So, using opinion mining and sentiment analysis methods [3] for populating ontology with subjective information would be interesting.

On the other hand, it would be possible to use different approaches for mapping and integrating ontologies and databases [9,15] in order to query any relational data base without needing to know anything about its structure.

Furthermore, our natural language interface makes the relation between humans and the Semantic Web easier. The methodology presented in this work has some drawbacks. First, our method depends on the label annotations of each class and instance. The ontology used in the experiment has full narrative labels, but we cannot guarantee that all the knowledge entities in the queried ontologies are correctly annotated. This is contrary to the recommendations of methods for evaluating ontology quality in reuse, such as ONTOMETRIC [16], which recommends assigning a correct natural language description to each knowledge entity in the ontology. Second, the systems only considers grammatically correct sentences. We are planning to include a spellchecker [24] and a grammar checker in the methodology, so that incorrect ones might be corrected and used.

Our approach is centred on obtaining individuals or the value of some attributes of the individuals. We are planning to extend the method to ask for classes, subclasses and others ontological elements. In addition, queries involving reasoning capabilities, like "If I finish my visit to Louvre at 5 pm on Monday, shall I visit Pompidou Museum on the same day?", will be examined and made available.

Acknowledgements

We thank the following institutions for their respective financial support: the Spanish Government under project TSI2007-66575-C02-02, the Government of Murcia under project TIC-INF 06/01-0002. We also thank the Tourism Department of the Region of Murcia for providing the tourism-related text.

References

1. Jena. semantic web framework (2007)
2. Baker, C.J.O., Su, X., Butler, G., Haarslev, V.: Ontoligent interactive query tool. Canadian Semantic Web 2, 155–169 (2006); PT: S; CT: Canadina Semantic Web Working Symposium (CSWWS 2006); CY: JUN 06, 2006; CL: Quebec City, CANADA; J9: SEMANTIC WEB BEYOND; GA: BEN54
3. Balahur, A., Montoyo, A.: Determining the semantic orientation of opinions of products- a comparative analysis. Procesamiento del lenguaje natural 41, 201–208 (2008)
4. Bernstein, A., Kaufmann, E.: Gino-a guided input natural language ontology editor. In: Cruz, I., Decker, S., Allemang, D., Preist, C., Schwabe, D., Mika, P., Uschold, M., Aroyo, L.M. (eds.) ISWC 2006. LNCS, vol. 4273, pp. 144–157. Springer, Heidelberg (2006)
5. Cardoso, J.: E-tourism: Creating dynamic packages using semantic web processes. In: W3C Workshop on Frameworks for Semantics in Web Services. IMPRESO (2005)
6. Delahousse, J.: Semantic web use case: An application for sustainable tourism development. IMPRESO (2003)
7. Dittenbach, M., Merkl, D., Berger, H.: A natural language query interface for tourism information. In: 10th International Conference on Information Technologies in Tourism (ENTER 2003), Helsinki, Finland, pp. 152–162 (2003)
8. Escamez, F., Beviá, R.I., Escamez, S.F., González, V., Vicedo, J.L.: An user-centred ontology-and entailment-based question answering system (in spanish). Procesamiento del lenguaje natural 41, 47–54 (2008)

9. Euzenat, J., Shvaiko, P.: Ontology Matching. Springer, New York (2007)
10. Fikes, R., Hayes, P., Horrocks, I.: Owl-ql a language for deductive query answering on the semantic web. Web Semantics: Science, Services and Agents on the World Wide Web 2(1), 19–29 (2004)
11. Jakkilinki, R., Georgievski, M., Sharda, N.: Connecting destinations with ontology-based e-tourism planner. In: 14th annual conference of the International Federation for IT&Travel and Tourism, January 2007. IMPRESO (2007)
12. Kanellopoulos, D.N., Kotsiantis, S., Pintelas, P.: Intelligent knowledge management for the travel domain. GESTS International Transactions on Computer Science an Engineering 30(1), 95–106 (2006)
13. Karvounarakis, G., Alexaki, V., Christophides, V., Plexousakis, D., Scholl, M.: Rql: A declarative query language for RDF. In: Proc. 11th Int'l Word Wide Web Conf. (WWW 2002), pp. 592–603 (2002)
14. Knublauch, H.: Travel.owl (2004), http://protege.stanford.edu
15. Konstantinou, N., Spanos, D.E., Chalas, M., Solidakis, E., Mitrou, N.: Visavis: An approach to an intermediate layer between ontologies and relational database contents. In: Proceedings of Workshops and Doctoral Consortium, The 18th International Conference on Advanced Information Systems Engineering - Trusted Information Systems (CAiSE 2006), pp. 1050–1061 (2006)
16. Lozano-Tello, A., Gmez-Prez, A.: Ontometric: A method to choose the appropriate ontology. Journal of Database Management 15(2), 1–18 (2004)
17. Missikof, M., Werthner, H., Hpken, W., Dell'Erba, M., Fodor, O., Formica, A., Taglino, F.: Harmonise towards interoperability in the tourism domain. In: 10th International Conference on Information Technologies in Tourism (ENTER 2003), pp. 29–31 (2003)
18. Natural Language Processing Group. University of Sheffield GATE. General Architecture for Text Engineering (2007)
19. Prantner, K.: Ontour: The ontology. In: Deri Insbruck (2004)
20. Prud'hommeaux, E., Seaborne, A.: Sparql query language for RDF, w3c candidate recommendation. World Wide Web Consortium (2006)
21. Salton, G., McGill, M.J.: Introduction to Modern Information Retrieval. McGraw-Hill, Inc., New York (1986)
22. Seaborne, A.: Rdql-a query language for RDF. W3C Member Submission 9, 29–21 (2004)
23. van Heijst, G., Schreiber, A.T., Wielinga, B.J.: Using explicit ontologies in kbs development. International Journal of Human-Computer Studies 46(2/3), 183–292 (1997)
24. Vilela, R.: Webjspell an online morphological analyser and spell checker. In: Proc. 11th Int'l Word Wide Web Conf (WWW 2002), vol. 39, pp. 291–292 (2007)
25. WTO. Thesaurus on Tourism and Leisure Activities of the World Tourism Organization (2001)

ItemSpider: Social Networking Service That Extracts Personal Character from Individual's Book Information

Tetsuya Tsukamoto[1], Hiroyuki Nishiyama[2], and Hayato Ohwada[2]

[1] Graduate School of Science and Technology, Tokyo University of Science
Noda, Chiba, 278-8510, Japan
[2] Research Institute for Science and Technology
j7407618@ed.noda.tus.ac.jp, {nisiyama,ohwada}@ia.noda.tus.ac.jp

Abstract. In this study, we develop a social networking service that extracts personal character from a user's book information. This system can share a user's personal book collection with other users and support the formation of a new community by providing the user's friends with a link to his/her book information. We define the user attribute as a vector comprising the book's details (e.g., author and publisher). For a set of users having the same books, a user clustering experiment was conducted on the basis of the user attribute value. In the experiment, users having the same books were observed to be classified in a group.

1 Introduction

Social networking service (SNS) is a service that constructs social networks. This service serves to establish and maintain contact between people. It offers a place for communication with close friends, acquaintances, or any person with similar interests. Since the SNS is widely used all over the world, it holds a considerable amount of user information. However, the SNS has little information for making a decision. That is general demographic data which users wrote when they started up. Hence, it is difficult to find a new person with interests similar to those of an existing user. The main reason for this difficulty appears to lie in the fact that there are few functions for recommending another person to a user in the SNS.

Recently, the number of collection management services has increased. These services manage books of individuals on the Web. These services obtain data on books by using the Amazon Web Service (AWS), which offers information on books, and build book collections. A user can browse the book collection by accessing these services in which many people contribute to comprise book information. There are few examples of extension that utilize book information between users. A study has previously been performed to understand the correlation between user's behavior in a social network and their characteristics. In order to facilitate the use of book collection as a user feature, it is important to clarify the relation between the SNS and users by performing data mining.

The purpose of this study is to develop the system named ItemSpider that integrates the social network on SNS with user's book collections to improve

D. Richards and B.-H. Kang (Eds.): PKAW 2008, LNAI 5465, pp. 161–172, 2009.

the service provided by the SNS and collection management services. As is increasing the amount of user's book information, ItemSpider is able to accurately characterize the users by extracting user interests from the book collection. If it is possible to compare and classify the characterized users in our system, we would be a step closer to finding a new user community.

The rest of this paper is organized as follows. In Section 2, we present related studies, and in Section 3, we describe the ItemSpider configuration. In Sections 4 and 5, we specify the functions of ItemSpider; the experimental results are presented in Section 6. We conclude in Section 7 by discussing our future plans.

2   Related Studies

In this section, we present some systems related to book management services and studies that pertain to user prediction using user behavior on the Web system.

Systems that manage books are of two types. The first is a system in which the user collects books in his/her own PC [7]. The second is a system that is constructed on the Web and is provided as a service [8]. Recently, new services of the latter type were launched. These systems mainly use the AWS to obtain book information, and provide unique functions (e.g., special view of individual book collections and associative searching). However, from the viewpoint of data mining, e.g., for determining similar user interests and locating the person with a specific book, little attention has been paid to the application of the collected data.

Book management services focus on personalization, which offers information to the user on the basis of his/her behavior. Personalization predicts a user's features from the log data of his/her behavior (e.g., search query, search result, and purchase history) and obtains information that suits a user. The personalized search [9] facility of Google and the product recommendation facility on e-commerce sites such as amazon.com are typical examples of personalization. By saving the search queries and archived pages of a user who has a Google account for each Google service, Google employs them when providing results for the next search operation. In amazon.com, an algorithm [3] that performs extended collaborative filtering for providing product recommendations is employed. As stated above, trends on user behavior obtained on the Web site are used for improving the service of the Web site.

Studies concerning predictions of the user details from his/her behavior on the Web are popular. Reference [1] predicts demographic data of users browsing the Web, the study in [4] relates to software that extracts a user's opinions from his/her weblog, and the study in [2] evaluates the most appropriate advertisement for a user from his/her behavior. Large-scale experimental results included in the study in [6] indicate that there is a correlation between the trend of a user's search and the demographic data in the MSN Messenger Network.

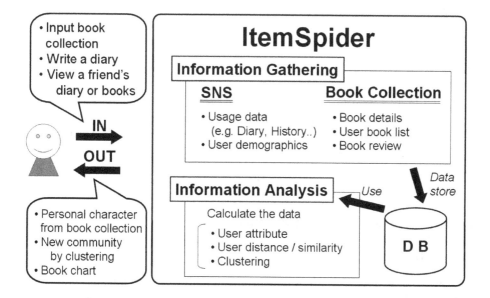

Fig. 1. Design of ItemSpider

3   ItemSpider

In this study, we develop a system named ItemSpider that integrates the SNS with a book collection service. This system considers the user's book collection as a user attributes for characterizing the user, and our purpose is to calculate the similarity between users by using their attributes. We aim to find users with similar interests.

Users write a diary, view a friend's diary or books, and input book collection in ItemSpider via ItemSpider viewers. Users finally get personal character, new community and book chart graph derived from ItemSpider functions. These functions are shown in Fig. 1. the functions are classified into "Information Gathering" and "Information Analysis".

The Information Gathering part gathers user behavior data from the SNS and his/her book collection to determine the user's features. It has an interface that allows this information to be displayed for public view.

1. **SNS:** It obtains the user's demographics and SNS usage data from a *user information admin module* and a *communication module* (shown in Section 4).
2. **Book Collection:** It has two modules to register data on books. The first is to automatically obtain detailed information on books from the AWS. The second is to add additional data that are manually put by users.

The Information Analysis part defines the user attributes and the similarity between users. Then it determines the features of the user on the basis of the data gathered by ItemSpider. Finally, it generates a new community from the features of the users by performing data mining. The functions/operations that belong to this part are as follows.

1. **Feature Extraction:** It extracts user attributes from the user's book collection.
2. **Calculating Similarity:** It is based on vector representation of a user attribute to calculate the similarity between two users.
3. **Clustering:** It is performed to identify similar users and group them on the basis of user attributes.

In the following Section, we describe these functions in detail.

4   Information Gathering Part

4.1   SNS

We employed the *user information admin module* and the *communication module* for constructing the social network. By using these modules, the system tracks the Web usage to obtain the user's features and behavior.

In the *user information admin module*, a user can configure the user profile and administration of access to the user's Web site. The system tracks the user behavior log by recording login time and page transitions.

In the *communication module*, the system offers functions that support the transmission of the user's information (e.g., the function of writing a diary and sending and receiving messages) and monitors the user behavior log by tracking comments to friends' diaries and the message history. A user can obtain a friend's diary, message, and recent information on books, reviews, book chains, etc.

4.2   Book Registering Module

A user has two methods to register a book. The first is to input an ISBN or a list of ISBN. (The ISBN stands for International Standard Book Number. It is unique, numerical, and commercial book identifier.) The second is to search for the book using a keyword. Fig. 2 shows the flow diagram of the process for registering a book in the system.

In the first method, ItemSpider obtains the ISBN from the user in a *take book information module* and then connects to the AWS which searches for book information using the ISBN. Finally, ItemSpider receives book information from AWS and registers it in the database.

In the second method, ItemSpider connects to the AWS which searches for the book list using the keyword given by user. ItemSpider then shows the user the search result set. If the user finds the desired book in the search result set, the associated ISBN is obtained and sent to *take book information module*. Book information is then added to his/her book collection. Fig. 3 shows the flow diagram illustrating the two input methods and the system interfaces.

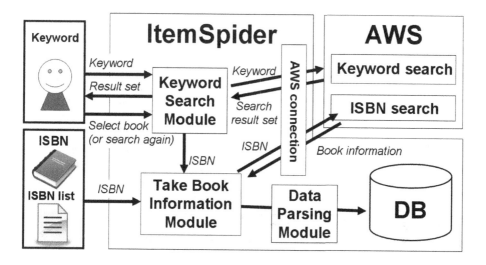

Fig. 2. Flow diagram illustrating the two input methods

Table 1. Details of the book information (the item with an asterisk has more than one element)

AWS information	book information			
book details	ISBN-10	ISBN-13	Title	ASIN
	* Author	* Author-role	* Other data	Size
	Publish-date	Publisher	Binding	Price
	Cover picture S/M/L		Amazon independent URL	
	Amazon product group			
related product	* Associated book			
category	* Category			

The data obtained from the AWS comprises three types of information: book details, related products, and categories. The response of the AWS is in XML format, and therefore, ItemSpider stores the data in the database by using data parsing module (that extracts specific data tags from the response). Table 1 shows the details of the data stored in the database.

4.3   Extra Data Addition Module

In addition to book detail data shown in Table 1, we introduce other types of data that are manually input by a user, such types of data are classified as follows.

Book Shelf: It reflects a bookshelf in the real world and categorizes user books from his/her desired viewpoint. A user can append data on one shelf per book

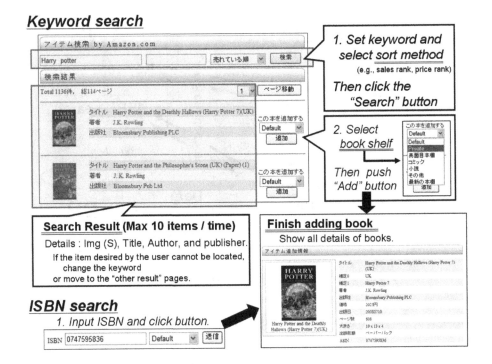

Fig. 3. The image of system from illustrating the two input methods and the system interfaces

and specify whether access to each bookshelf is open to the public or to specific users.

Book Review: It registers a book review and a grade on the five-grade evaluation scale. This review will be usable for discriminating community. It is also accessible to the public.

Book Chain: It reflects a list of books that are grouped together with respect to user desired viewpoints. It differs from book shelf in that a single book falls into several book categories. It is also accessible to the public.

Book Title Attribute: Book information obtained from AWS includes the true title and its complementary information such as version and series. Fig. 4 shows a typical title information. This includes the true title ("Harry Potter and Deathly Hallows") and its complementary information ("Harry Potter 7", "UK" and "Adult edition"). We adopt title attribute and keyword attribute as a book attribute, a specific tag is attached to each information; a title tag is attached to true title, and keyword tags to the complementary information as shown in Fig. 4. Here, keyword tagged information can also be added manually by users so as to enrich the book information.

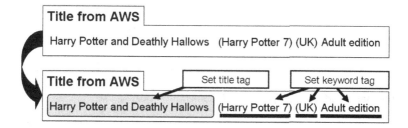

Fig. 4. Example of a user-set keyword in the title obtained from the AWS

4.4 Output

The user manages his/her book collection and obtains the compiled or mined data through a browser. Fig. 5 shows the top page of ItemSpider and Fig. 6 shows an interface for viewing the book. On the viewer, it is possible to change view type, book category and sort, and to browse the user's book collection refining to the search result to the book list user wants. It is also possible to browse book collections of other users.

The system provides a summary of the data on the following fields: book category, author, publisher, price, published date, and the date when the user added the book to the system. The five former fields have a hierarchical structure to organize products; this structure is provided by AWS as the *BrowseNode* information. In this case, we focus on a third-level of node in *BrowseNode* for well-balanced classification of book products. ItemSpider counts the number of books falling into a given set of book categories, and a user can browse the data summary that presents every bookshelf in the form of graphs or numbers.

5 Information Analysis Part

In ordinary SNS, there is a user community and all users are connected. In this study, the system creates new communities by using the book collection that it possesses. First, the data that the system collects from book collection as feature is added to every user. Subsequently, data mining is performed to obtain user features. Thereby, groups are created in the system. It can be presumed that a user can find the books collected by other users in his/her group, which will enable him/her to discover persons with similar interests.

In the following sections, user attributes, similarity between two users, and data mining techniques used in our system are described.

5.1 User Attribute

The user attributes include six summary data that are described in Section 4.4 and the demographic data input by the user. In this paper, we explain the user attributes that are limited to 27 categories from summary data. Table 2 lists all the categories.

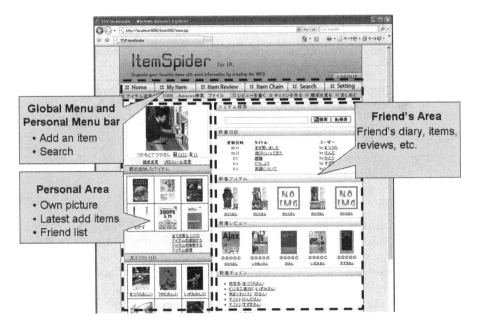

Fig. 5. Top page of ItemSpider

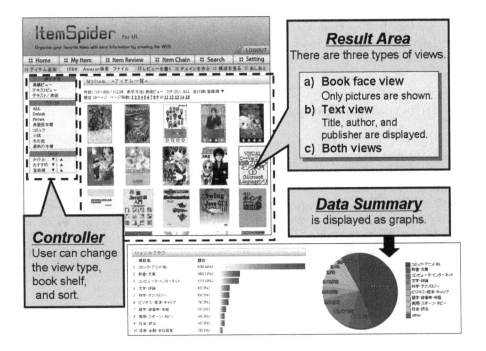

Fig. 6. Interface of viewer and data summary

Table 2. Category list contained in category summary data

Magazine	Entertainment	Talent photography	LanguageEYearbook
Old book	Qualification	HistoryEGeography	PocketbookELibrary
Child	ArtEDesign	SocietyEPolitics	EducationEExamination
Adult	LifeEHealth	ScienceETechnology	ComputerEInternet
Nonfiction	SportEHobby	LiteratureECriticism	InvestmentEFinance
Travel guide	ScoreEMusic	HumanitiesEThought	BusinessEEconomy
Game manual	ComicEAnimation	MedicineEPharmacy	

5.2 Distance and Similarity

ItemSpider calculates the distance between two users by using the Euclidean distance technique and determines the similarity between them by using the vector space method [5]. Let a user feature vector of user i for a particular book summary data be denoted by A_i. Further, let a_{ik} be the value of user feature k of user i, where $1 \leq k \leq m$ and $A_i = (a_{i1}, \cdots, a_{im})'$. The distance between user i and user j, d_{ij}, is expressed as

$$d_{ij} = \sqrt{\sum_{k=1}^{m} \mid a_{ik} - a_{jk} \mid^2} \qquad (1)$$

The similarity between user i and user j, S_{ij}, is described as

$$S_{ij} = \frac{A_i \cdot A_j}{\mid A_i \mid\mid A_j \mid} \qquad (2)$$

5.3 Clustering User Attributes

We perform clustering for forming groups based on the user attributes; we use WEKA [11], which is a well-known data mining tool written in Java. Thus, we obtain a user group where every user has a high degree of similarity with the other users and extract the features of the group. The simple K-means algorithm is used for the grouping. All users are classified to the group that includes users who have median point of category vector close to each other.

6   Experiment

In the experiment performed in this study, there were a small number of real users in ItemSpider. Therefore, we collected sample users who input a book collection similar to the real users. We investigate the relationship between the grouping of users and the similarity of their book collections. We use overlap ratio of user's book collections for measuring the similarity between users.

The experiment compares the collections of the real users and sample users. The latter have a book collection with an overlap on real user's book collection

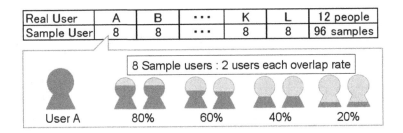

| Real User | A | B | $\cdots$ | K | L | 12 people |
| Sample User | 8 | 8 | $\cdots$ | 8 | 8 | 96 samples |

8 Sample users : 2 users each overlap rate

User A   80%   60%   40%   20%

Fig. 7. Composition of real and sample users in this experiment

in four kinds of a percentage (20%, 40%, 60%, and 80%). In the experiment, we have two sample users per percentage pattern as shown Fig. 7. The data of the real and sample users clarify the manner in which the overlap percentage influences the grouping. The total number of real users is 12 and the total number of sample users is 96.

We executed WEKA for the experiment and WEKA produce ten clusters. For each real user, we manually calculate the similarity between the real user and corresponding sample users in every cases where the sample users are in the same cluster and the different clusters. We then counted the number of sample users belonging to the same cluster for the real user. Table 3 shows the total sum of same cluster rate for each overlap ratio. The table also shows the similarities of the same cluster users and those of the different cluster users in average. Fig. 8 illustrates how to calculate the same cluster rate and the similarity between users at 40% overlap ratio.

6.1 Discussion

From Table 3, we find that the degree of similarity increases with the overlap ratio, because sample users who have a high overlap ratio of book collection might be more similar to real users. The table indicates that the similarity of same cluster is higher than that of different clusters at every overlap ratios, though there are few differences between the real user and the sample user both when they are in the same group and when they are in different groups. In addition, the sample user who has a high degree of similarity to the real user is not always in the same group as the real user. For example, user pairs are not classified in the same group despite their similarity being close to 100%.

The above results reveal new facts. The first is that the clustering technique does not allow groups to overlap. This means that two users tend to be classified into distinct groups, even though they have almost the same book collection. Another fact is that a count-based feature vector representation may ignore a user's preference. Since this will obstruct the system from detecting user features, more appropriate user feature representation should be explored.

It should be mentioned that we used *BrowseNode* in AWS as book category data. A point to be noted is that the system detects only a part of the user

Table 3. Result of the classification experiment

Overlap ratio	Same cluster rate	Similarities between users		
		Average	Same cluster	Different clusters
20%	3/24	0.8716	0.9024	0.8672
40%	10/24	0.9343	0.9367	0.9330
60%	14/24	0.9666	0.9682	0.9616
80%	19/24	0.9875	0.9883	0.9861

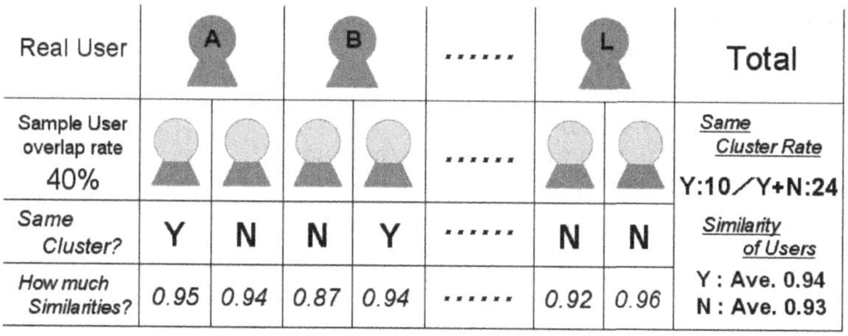

Fig. 8. How to calculate the same cluster rate and the similarity between users at 40% overlap ratio

features (including the book category) because the amount of information in *BrowseNode* varies from one book to another. The system may need to add new information apart from *BrowseNode* to the book.

7   Conclusion

In this paper, we describe the design and functions of ItemSpider, which extracts personal details from his/her book information. We conduct an experiment on grouping users with similar book collections. It is observed that the degree of similarity between users increases with the overlap ratio. Finally, we discuss issues based on user attributes that are involved in the grouping process.

In future, we are going to publish ItemSpider for the student of the same department in our university, and gather live data from usage history. We shall also propose a type of grouping by using many kinds of user attributes from actual data and utilize grouping issues in Section 6. It is hoped that the present study will contribute realizing intelligent SNS.

References

1. Adar, E., Weld, D.S., Bershad, B.N., Gribble, S.S.: Why We Search: Visualizing and Predicting User Behavior. In: Proceedings of the 16th international Conference on World Wide Web, May 08-12 (2007)

2. Hu, J., Zeng, H.-J., Li, H., Niu, C., Chen, Z.: Demographic Prediction Based on User's Browsing Behavior. In: Proceedings of the 16th International Conference on World Wide Web, May 08-12 (2007)
3. Linden, G., Smith, B., York, J.: Amazon.com Recommendations: Item-to-Item Collaborative Filtering. IEEE Internet Computing 07, 76–80 (2003)
4. Mei, Q., Ling, X., Wondra, M., Su, H., Zhai, C.: Topic Sentiment Mixture: Modeling Facets and Opinions in Weblogs. In: Proceedings of the 16th International Conference on World Wide Web, May 08-12 (2007)
5. Salton, G., McGill, M.J.: Introduction to Modern Information Retrieval. McGraw-Hill, New York (1983)
6. Singla, P., Richardson, M.: Yes, There is a Correlation: From Social Networks to Personal Behavior on the Web. In: Proceeding of the 17th International Conference on World Wide Web, April 21-25 (2008)
7. Delicious Library, http://www.delicious-monster.com/
8. Booklog, http://booklog.jp/
9. Google Account, http://www.google.com/accounts/
10. Amazon Web Service, http://aws.amazon.com/
11. WEKA, http://www.cs.waikato.ac.nz/ml/weka/

Acquiring Marketing Knowledge from Internet Bulletin Boards

Hiroshi Uehara and Kenichi Yoshida

Graduate School of Business Science, University of Tsukuba

Abstract. This paper proposes a method to acquire marketing knowledge from Internet bulletin boards. It aims to clarify how viewers' interest in TV advertisements are reflected on their perceived images on the promoted products. By analyzing viewers' interests, we can acquire various knowledge which we can use to promote products. Two kinds of time series data are generated based on the proposed method. The first one represents the time series fluctuation of the interest in the TV advertisements. The other one represents the time series fluctuation of the image perception on the products. By analysing the correlations between these two time series data, we will try to clarify the implicit relationship between the viewer's interest in the TV advertisement and their perceived image of the promoted products. By applying the proposed method to an Internet bulletin board that deals with certain cosmetic brands, we will show how we can acquire marketing knowledge from Internet bulletin boards.

1 Introduction

The rapid popularization of online communities, such as those formed using Internet bulletin boards, blogs, and SNS, has increased their importance as a significant source of marketing knowledge [Gruhl 04, Haga 02]. Marketers have to pay careful attention to the opinions formed in online communities. Thus, knowledge acquisition from online communities has to be investigated.

This paper proposes a method to acquire marketing knowledge from Internet bulletin boards. It aims to clarify how viewers' Interest in TV advertisements are reflected on their perceived images of the promoted products. By analyzing viewers' interests, we can acquire various knowledge by which we can use to promote products. Two kinds of time series data are generated based on the proposed method. The first one represents the time series fluctuation of the interest in the TV advertisements. The other one represents the time series fluctuation of the images on the products. By analysing the correlations between these two time series data, we will try to clarify the implicit relationship between the viewer's interest in the TV advertisement and their perceived images of the promoted products. By applying the proposed method to an Internet bulletin board that deals with certain cosmetic brands, we will show how we can acquire marketing knowledge from Internet bulletin boards.

D. Richards and B.-H. Kang (Eds.): PKAW 2008, LNAI 5465, pp. 173–182, 2009.

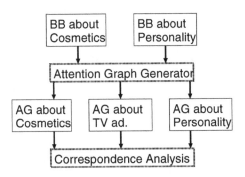

Fig. 1. Outline of Proposed Method

In this paper, Section 2 explains the basic ideas, and Section 3 reports on the example analysis. Finally, Section 4 concludes our findings.

2 Correspondence Analysis of Internet Bulletin Boards

2.1 Outline of Proposed Method

Fig. 1 shows the outline of the proposed method. It first extracts "attention graphs" [Uehara 05] from Internet bulletin boards. The attention graph shows the time series fluctuation of the opinions formed in online communities. It sometimes shows the viewers' interest in the TV advertisements, certain products, and personalities of TV advertisements.

In the example analysis shown in the next section, we used two Internet bulletin boards. The first bulletin board (Fig. 1 "BB about Cosmetics") is a collection of various opinions on cosmetic brands. The community is mainly formed by young ladies who actually used the cosmetics discussed. They are exchanging various opinions on products and TV advertisements. By analysing their comments, we can generate an attention graph about their product image perception (Fig. 1 "AG about Cosmetics") and an opinion of the TV advertisements (Fig. 1 "AG about TV advertisements"). We also used a second bulletin board about personalities (Fig. 1 "BB about Personality"). From this bulletin board, we could generate an attention graph about personalities' images (Fig. 1 "AG about Personality").

The proposed method then performs a correspondence analysis [Benzecri 92] using the generated attention graphs. Correspondence analysis is a typical method used by marketers. Although it is typically used in questionnaire surveys [Aaker 81, Schlinger 79, Wells 71],the application of correspondence analysis to the attention graph allows for the opinion of online communities to be analysed. Whereas, questionnaire surveys have difficulty in analysing the rapid change of opinions frequently observed in online communities, the direct use of Internet bulletin boards contributes to the rapid analysis of such changes in opinion.

The example analysis shown in the next section shows how TV advertisements change the product image of online communities.

2.2   Algorithm for Generating Attention Graphs

The attention graph creation algorithm is originally proposed in [Uehara 05]. Fig. 2 shows its outline. It inputs two files. "TimeStampedWordsFile", the output file from the pre-processing component, has the list of morpholized words from the Internet bulletin boards. Each word in this file has a time stamp derived from the Internet bulletin boards, thus all the words are sorted by the time sequence. Another file is KeyWordFile. This file includes the same words as TimeStampedWordsFile. The differences are that the words in this file are attached with a threshold on the interval, and each word is uniquely registered.

The algorithm first reads a word from KeyWordFile then, searches for the same word in TimeStampedWordsFile. Every time the word is found, the algorithm calculates the interval between the time stamp of the word and the time stamp of the one word previously found. If the interval is less than the threshold, the algorithm tentatively stores the word with its time stamp in WkAttnGraphRec then increments the counter. On the other hand, if the interval surpasses the threshold, the algorithm checks if the counter of the word in WkAttnGraphRec surpasses the threshold of the frequency. If it is true, stored data in WkAttnGraphRec is recognized as symbolic words, and written in AttnGraphFile. If it is false, the stored data is cleared out. By decrementing the threshold from the given threshold down to 1, this algorithm extracts every period which represents different degrees of viewers' enthusiasm. Here the periods share the same symbolic word. With a higher threshold, periods of higher enthusiasm are extracted. With a smaller threshold, periods of lower enthusiasm are extracted. After that, the next identical word is sought out in TimeStampedWordsFile. If it is found, the threshold is reset back to the given value. Then the algorithm again extracts the periods of viewers' attention. This procedure is repeated by decrementing the threshold until it reaches 1.

When TimeStampedWordsFile reaches its end, the algorithm reads the next word from KeyWordFile, and starts the procedure above all over again starting from the first record in TimeStampedWordsFile to the end. Thus, all the data for attention graphs are written in AttnGraphFile.

2.3   Correspondence Analysis of Attention Graphs

Attention graphs about TV advertisements are based on a frequency of personality names appearing in Internet bulletin boards while the TV advertisements are being broadcasted. It reflects the strength of the impression formed by the TV advertisements. By making a correspondence plot (See next section, [Benzecri 92]), we can clearly show the impact of TV advertisements. Attention graphs about product image are based on a frequency of "image word" appearing in Internet bulletin boards. It reflects the product image formed by the

Main Procedure createAttentionGraphs
Input
TimeStampedWordsFile : Word list with time stamps
KeyWordFile : Words list with the threshold "$Th2_w$"

Output
AttnGraphFile ; Attention Graph data comprised of Symbolic
Words , durations, Attention Levels
Var
WkAttnGraphRec ; Work record tentatively storing
information for Attention Graph

Begin
Th1 = threshold of consecutive appearance of w

while (Read a word and its $Th2_w$ from KeyWordFile) {
// Every time when a word is picked out from KeyWordFile,
reading procedure of TimeStampedWordFile starts over
from the first record.
while (read a word from TimeStampedWordsFile){
for (decrement *Th2w* until it reaches to 1){

if (the word from KeyWordFile matches to the
word from the word from TimeStampedWordsFile
& interval between this match and last match is
less than $Th2_w$) {
update ending time of Attention Graph in
WkAttnGraphRec, and increment the number
of matches
}

else if(the word from KeyWordFile matches to the
word from the word from TimeStampedWordsFile
& interval between this match and last match is
greater than Th2w) {
if (The number of matches greater than *Th1*) {
write WkAttnGraphRec to AttnGraphFile
}
set time stamp from TimeStampedWordsFile
to WkAttnGraphRec as the starting time of
Attention Graph
set a word from TimeStampedWordsFile to
WkAttnGraphRec as a Symbolic Word
}
}
}
Write all the data remaining in WkAttnGraphRec to
AttnGraphFile when reading procedure of
TimeStampedWordsFile reaches to the end

}
End

Fig. 2. Algorithm for Generating Attention Graphs

TV advertisement. By making a correspondence plot, we can show the product image formed by the TV advertisements.

We can also analyse the personality images using the attention graphs generated from the Internet bulletin board about personalities.

3   Example: Analysis of Cosmetic Brand Advertisement

To show the ability of the proposed method, we have analysed the effect of TV advertisements for a certain cosmetic brand. The series of TV advertisement uses 4 personalities, and tries to portray a good image of the products. From a period of August 2005 to October 2006, we analysed two Internet bulletin boards. The first bulletin board is a collection of various opinions about cosmetic brands.

Fig. 3 shows the attention graph generated. The horizontal axis is a period of TV advertisements, and the vertical axis shows the strength of the impact of the TV advertisements. The successive TV advertisements show a higher impact of its personalities, and this graph enables the comparison of the impact of TV

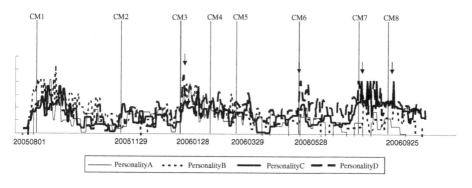

Fig. 3. Attention Graph about "TV Advertisement"

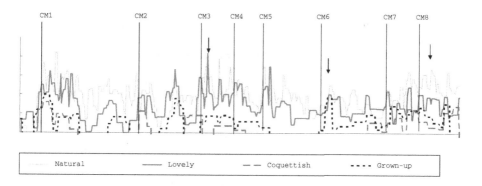

Fig. 4. Attention Graph about "Product Image"

Table 1. Image Vectors

(a) Image Vector of TV advertisement

Personality	Attractive CMs					other CMs		
	CM1	CM3	CM6	CM7	CM8	CM2	CM4	CM5
A	17	16	4	1	2	3	5	4
B	38	21	5	4	5	5	7	5
C	14	11	5	21	13	5	5	6
D	12	22	21	24	11	5	9	8

(b) Image Vector of Product

Image words	Attractive CMs					other CMs		
	CM1	CM3	CM6	CM7	CM8	CM2	CM4	CM5
Natural	24	22	15	13	25	27	19	21
Colorful	4	5	4	5	3	0	4	4
Elegant	8	7	6	7	4	8	8	4
Tough	3	3	0	2	3	0	2	2
Lovely	12	17	13	9	8	14	17	17
Coquettish	3	1	0	4	3	5	2	0
Deep	1	0	0	6	4	0	2	0
Great	3	1	2	0	3	0	0	0
Grown-up	5	3	6	5	4	0	4	2
Dark	4	4	11	13	6	11	6	11
Light	3	3	9	4	3	3	8	4
Bright	3	3	2	4	4	3	2	4
Calm	3	0	0	0	4	0	2	2
Beautiful	22	33	32	28	25	30	25	30

advertisements. From this graph, we can acquire various knowledge about online communities. For example:

- CM1, CM3, CM6, CM7 and CM8 could successfully draw attention from online communities. But attention to CM2, CM4 and CM5 are not very significant.
- Personality D drew the most attention during CM6.

Fig. 4 shows the attention graph on a product image.[1] We can see the image formed by the TV advertisements from this graph.

Table 1 are the table format data converted from attention graphs. Each column shows the average impact during the broadcast of TV advertisement. From this table format data, we can generate correspondence plots (See Fig. 5, 6, 7).

---

[1] Since the attention graph about product image involves frequencies about various image words, Fig. 4 shows only the part of them which improve readability. Fig. 6, 7 and Table 1 also shows the partial information.

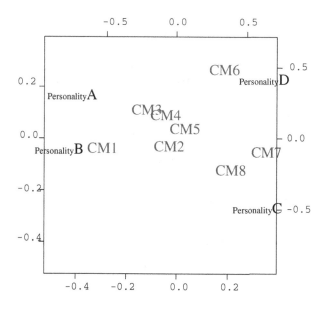

Fig. 5. Correspondence plot of "TV Advertisement"

Fig. 5 shows the correspondence plot of "TV Advertisement". As shown in the figure, Personality B drew most attention during CM1. This implies that the image of CM1, i.e. the image of the product treated in CM1, is affected by the image of the personality B. Personality D drew most attention during CM6, and Personality C drew most attention during CM7 and CM8. The impact of Personality A is not clear from this analysis. Fig. 6 shows the correspondence plot of "Product Image". As shown in the figure, the image of CM1 is "Natural", "Great, and "Tough". The image of CM3 is "Lovely", and the image of CM8 is "Cheerful". Fig. 7 shows the correspondence plot of "Personality Image". The image of personality B is lovely and big-eyes, and the image of personality C is elegant.

By comparing the correspondence plots, we can see:

- Both CM3 and personality B have the image of "Lovely" and "Beautiful". Since the change of product image follows the the frequency change of TV advertisements' impact, this implies that the image of personality, i.e. "Lovely" and "Beautiful", affects the product image treated in CM3.
- Similar phenomenon are observed between CM6 and personality D, and between CM7 and personality C, and etc.

Fig. 8 shows the autocorrelation of TV Advertisements' impacts. From this figure, we see that the effect of TV Advertisements continues for about 17 days. After this period, TV Advertisements seem to lose their effect.

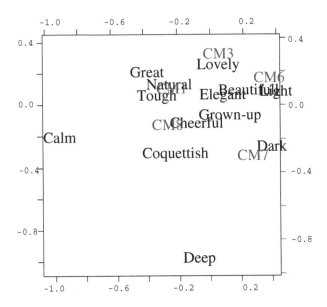

Fig. 6. Correspondence plot of "Product Image"

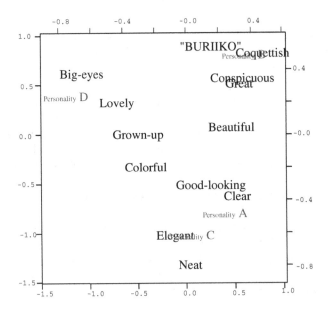

Fig. 7. Correspondence plot of "Personality Image"

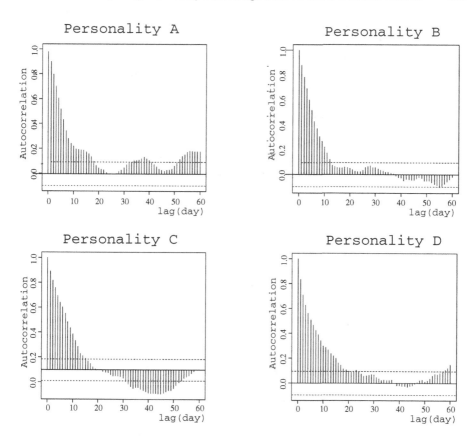

Fig. 8. Autocorrelation of "TV Advertisement"s Impact

4 Conclusions

This paper proposes a method to acquire marketing knowledge from Internet
bulletin boards. It aims to clarify how the viewers' interests on TV advertisements are reflected on their perceived images of the promoted products.

By applying the proposed method to an Internet bulletin board that deals with
certain cosmetic brands, we were able to show how we could acquire marketing
knowledge from Internet bulletin boards.

References

[Gruhl 04]   Gruhl, D., Guha, R., Liben-Nowell, D., Tomkins, A.: Information Dif-
                fusion Through Blogspace. In: Proceedings of WWW Conference, pp.
                491–501 (2004)
[Haga 02]   Haga, H.: Combining video and bulletin board systems in distance
                education systems. Elsevier Science, Internet for Higher Education 5,
                119–129 (2002)

[Uehara 05]   Uehara, H., Yoshida, K.: Annotating TV Drama based on Viewer Di-
                 alogue -Analysis of Viewers' Attention Generated on an Internet Bul-
                 letin Board. In: SAINT 2005, pp. 334–340 (2005)
[Benzecri 92]   Benzecri, J.-P.: Correspondence Analysis Handbook. Marcel Dekker,
                 New York (1992)
[Aaker 81]   Aaker, D., Bruzzone, E.: Viewer perceptions of prime-time television
                 advertising. Journal of Advertising Research 21, 15–23 (1981)
[Schlinger 79]   Schlinger, M.: A profile of responses to commercials. Journal of Ad-
                 vertising Research 19, 37–46 (1979)
[Wells 71]   Wells, D., Leavitt, C., McConville, M.: A reaction profile for TV com-
                 mercials. Journal of Advertising Research 11, 11–15 (1971)

A Design for Library Marketing System and Its Possible Applications

Toshiro Minami

Kyushu Institute of Information Sciences, Faculty of Management and Information
Sciences, 6-3-1 Saifu, Dazaifu, Fukuoka 818-0117 Japan
Kyushu University Library, Research and Development Division, 6-10-1 Hakozaki,
Higashi, Fukuoka 812-8581 Japan
minami@kiis.ac.jp, minami@lib.kyushu-u.ac.jp
www.kiis.ac.jp/~minami/

Abstract. Library marketing system is a system that helps with improving patrons' convenience and library management based on the data in libraries by analyzing them with data mining methods, including statistical ones. In this paper
we present a design of such a system which deals with usage data of materials
and extracts knowledge and tips that are useful, for example, for better arrange-
ments of bookshelves and for providing patrons with information which will attract the patrons. Two methods are proposed for collecting usage data from
bookshelves; one is with RFID and the other is with two-dimensional code, such
as the QR code that is very popularly used in mobile phones. By combining several analysis methods, we can construct a library marketing system, which will
give benefits to library management and patron services.

Keywords: Library Marketing, RFID (Radio Frequency Identification), 2-
dimensional Code, Intelligent Bookshelf, Data Mining.

1 Introduction

The aim of the system presented in this paper is to help libraries and librarians with
improving their patron services so that they can get better satisfaction from their patrons. The system also aims to directly provide the library patrons with useful information that helps them with choosing reading materials, finding useful information,
exchanging their views with other patrons, and so on.

In order to achieve these aims, the system collects data about the patrons and about
how they use the library materials. We consider two types of data collection methods
in this paper; with RFID and with 2-dimensional code. With RFID we collect usage
data of library materials and with 2-dimensional code we collect who is interested in
which books. The system then helps the librarians with analyzing these data and extracting various kinds of tips that are supposed to be helpful for better library management in collection of books, shelf arrangement, patron services, and so on. We will
call such a system a library marketing system.

D. Richards and B.-H. Kang (Eds.): PKAW 2008, LNAI 5465, pp. 183–197, 2009.

In this introductory section, we start with describing our motivation and the concept of library marketing in Section 1.1 followed by discussing its importance for libraries in Section 1.2.

1.1 Library Marketing

Marketing is one of the most important concepts for profit organizations in order to get good reputatons from customers and to have more profits. This concept is also important for non-profit organizations such as libraries, not in the sense of to increase profits but in the sense of to increase their good reputations and have better customer satisfaction (CS), or perhaps you may also call patron satisfaction (PS) as well.

There may be a lot of types of library marketing methods. In this paper we focus on the one using databases in a library, including log data of services provided by the library.

Figure 1 shows the basic structure for the library marketing in this type. It is the mutual depending structure between databases and services. In the right part of the figure are library services provided by the library. "OPAC" (Online Public Access Catalog) is a search engine for the library materials. "My Library" is a relatively new service that provides patrons with individual information such as which books they are borrowing and have been borrowed, a list of books the patrons might like to read, and so on. "Reference" means the online reference service. These are basic and at the same time very useful online library services.

These services depend on the databases that consist of catalog data, circulation data, patrons' profile data, etc. These data are collected at a circulation counter, at a self checkout machine, and in other ways. The log data, which are collected as the system provides various services to patrons, are also collected and are used as the service log database.

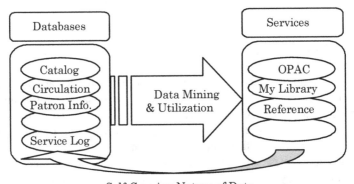

Self-Growing Nature of Data

Fig. 1. Library Marketing with Databases and Services

The collected data are supposed to be analyzed, or data mined, and the result should be reflected to the services in order to improve the quality of patron services. As the quality of services gets higher, more patrons will use the library and the system so that the system gets more service log data, and they can be used for better services for increasing CS, or PS. Such a natural self-growing mechanism is very important for library marketing with databases and services.

1.2 Importance of Library Marketing

Patron orientedness is very important for libraries and it is well described in "The five laws of library science" advocated by the famous Indian library scientist S. R. Ranganathan [10], which was published way back in 1960s. They are: (1) Books are for use, (2) Every reader his book, (3) Every book his reader, (4) Save the time of the reader, and (5) The library is a growing organism.

Even though these laws consider only the books and thus look old, they are still new and applicable even now as we rephrase them by replacing "book" with "information," "material," or "service." In order to be patron oriented, libraries have been growing, or changing, continuously by introducing up-to-date technologies.

For examples, libraries have introduced computers and started computerized processing for library jobs a couple of decades ago. They are one of the earlies organizations in computerization. With computer systems, libraries are able to provide OPAC service for book retrieval and connected the computers to the Internet so that their patrons are able to access to library services such as Web-OPAC, online referencing and other library services in 1990s. Now they are starting to provide the so called Web 2.0 services as new kind of library services. The SNS (Social Networking Service) is a typical example of such services.

Even now the most important service of libraries which we expect to libraries to provide is the one that helps the patrons with providing information materials, e.g. books, magazines, newspapers, etc. Suppose a patron is searching for a book and it is not in the collection of the library where he or she belongs to as a member, the library will try to find out other libraries that have the book and then it will choose and ask one the libraries for let it borrow the book for the patron. The book may be sent to the library for borrowing or a copy of a part of the book may be sent instead of the book itself. This is a rough sketch how the ILL (Inter-Library Loan) service is carried out.

Recently, not only the service of providing materials but also other services such as supporting the patrons with training their information literacy skills are considered to be very important for libraries.

Due to the advancement of network society, such library services are requested to be extended so that they are shifted to be more network-based, more collaborative with other libraries, hopefully providing 24 hours, and so on. In order to realize such services, libraries and library systems should change themselves to be more network-oriented and automated ones so that patrons are able to access to their libraries at any time from wherever they want.

As was illustrated in Figure 1, the network services suit to library marketing because the log data of these services are easy to collect in the system and these data are stored as a service log database.

The library services with physical materials such as printed books, printed magazines, CDs, DVDs, and etc. do not give data automatically. In a decade RFID (Radio Frequency Identification) [3] technology has been introduced to many libraries; the number is over 200 in Japan (by 2009). An RFID tag is attached to a physical material, e.g. a book, and an RFID reader can detect the tag's ID without contacting it. In this way the data relating to these materials can be automatically collected and will be stored in a database.

There is a report relating to library marketing, where they collect data on the numbers of patrons in a library's rooms together with entrance data and they propose a better space arrangement plan by analyzing the patrons' behavioral properties [4].

2 Tools for Library Marketing

As was described in the previous Section 1, we need some AIDC (Automatic Identification and Data Capture) technology [1] in order to collect data for physical materials. In this paper we deal with two types of technologies; RFID and 2-dimensional code. We will briefly describe the former one in Section 2.1 and the latter one in Section 2.2.

2.1 RFID and Intelligent Bookshelf

The principles of RFID tag system is illustrated in Figure 2(a). The RFID tag system consists of two major components; tags and reader/writers (R/Ws). A tag is able to communicate with a reader/writer when they are located sufficiently close each other.

As is shown in Figure 2(a) the RFID tag at the right-hand side consists of an IC chip and an antenna. It has no batteries and thus cannot run by itself. At the left-hand side is an R/W, which provides energy to the tag with its antenna. The RFID tag gets energy from the R/W with electro-magnetic induction via its antenna. It waits until sufficient energy is charged in it and when it is ready, it starts running and communicates with R/W and exchanges data such as its ID and status data by making use of the same antenna. There are two types of tags; one is read-only and the other is read-write. Normally the latter type is used in library application.

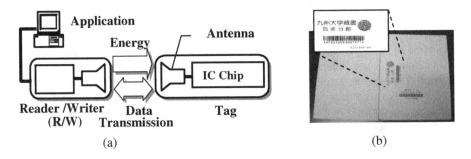

Fig. 2. (a) Principle of RFID (b) RFID Tag Attachment on a Book

Figure 2(b) shows an example of how RFID tag is attached on a book. It is an RFID tag used in Chikushi Branch Library of Kyushu University Library [5], Japan. The tag is built as a label on which the library name is marked together with the university logo. The material ID is also marked in barcode on the label. The barcode is supposed to be used when this material is carried to another library in the ILL program, i.e. for interoperability, and when the tag has been bad and does not respond to R/Ws for just in case.

Comparing to the barcode system which is mostly used in libraries now and is considered to be the standard technology to libraries, RFID tag system has an advantage that it is much easier to put material in an appropriate position. In order to read the book ID, you have to put the barcode part of a book in the appropriate area so that the barcode can be optically scanned by the scanner while with RFID tag all you have to do is to put the book in a designated area. As a result self checkout machine is easier to operate so that it is easy enough for children and elderly patrons to use. This is a very important difference.

So far the dominating reason for the libraries whey they introduce the RFID tag system is that it let the processes be more efficient; i.e. it is faster to proceed circulation, it is supposed to have less running cost, and thus the number of librarians needed will be smaller, etc., even though its initial cost is very high.

Typical usages of R/Ws in libraries are shown in Figure 3. They are security gate (Figure 3(a)), those for circulation counter (Figure 3(b)), self checkout machine (Figure 3(c)), and handy R/W for inventory (Figure 3(d)).

Application of RFID technology makes the following advantages:

(1) Efficient Checkout and Checkin

Checkout and checkin processing with RFID is much faster than with barcode. Furthermore use of RFID self-checkout and self-checkin machines will reduce the burden of library staff a lot as well. University of Nevada, Las Vegas library [12] has reduced the number of the circulation counters in half by adding self checkout/return machines.

(2) Easy to Use Checkout and Checkin

Furthermore, the self-checkout/checkin machines are much easier to use for patrons because the locationing restrictions for RFID tags are less than barcode tags, and because of the multiple reading with anti-collision mechanism of RFID that is impossible with barcode.

(a)   (b)   (c)   (d)

Fig. 3. RFID Reader/Writers for Libraries: (a) Security Gate, (b) Desktop Type for Circulation Counter, (c) Self Checkout Machine, and (d) Handy Type for Book Inventory

(3) Efficient Inventory
The inventory time reduces notably, probably from half to one tenth or even less. In a typical case service-closing period of the library for inventory changes from one to two weeks to a couple of days. City of Kita Central Library in Tokyo [2] spends about five to ten minitues for inventory every day, and finishes one cycle of inventory in a month for 300 thousand books.

An intelligent bookshelf (IBS) (Figure 4) is a bookshelf which has shelf type RFID R/W, where its RFID antennas are installed in the bookshelves so that the R/W can detect what books are put in which shelf in real time [6][7]. There are a couple of types of antennas. The book-end type is the one the antennas are put next to books like spacers as is in Figure 4. Another one is the shelf-board type, of which the antennas are put under the books either in between the books and the shelf-board or the boards themselves. Furthermore some IBS's antennas are put behind the books. For example if you use UHF type RFID tag system it is natural to put the anntennas behind the shelves so that they can be installed additionally on the popularly used bookshelves.

By using such equipments we can collect the usage data of books. By analyzing the usage data we may extract useful knowledge, i.e. library marketing [6][7][8][13]. For example, we can get the information how often a specific book is used and the differences of usage patterns according to the day of the week, time zone in a day, etc. We can provide such information to the library patrons. It may be used by library staff for planning improved services as well.

2.2 Two-Dimensional Codes and Mobile Phone

As was pointed out in Section 2.1, RFID technology has great advantages in material management. However not only the R/Ws but also the price of RFID tag is very high compared with that of barcode.

Fig. 4. Intelligent Bookshelf for Library Marketing

2-dimensional code is a technology inbetween RFID and barcode technologies in many aspects, thus it can be used as an alternative to RFID. Figure 5 shows a sample of barcode and 2-dimensional code (QR code). Other 2-dimensional codes include PDF417, DataMatrix, and Maxi Code [8]. 2-dimensional code has the following advantages in general.

(1) Low Cost
    Just like barcode, we can print 2-dimensional code with ordinary printers.

(2) Large Recordable Data Size
    The recordable, or printable, size of data is about tens of digits in barcode, while it is much large in 2-dimensional code system; for example some thousands digits or characters. In terms of recordable size it is competitive with RFID tags. A disadvantage of 2-dimensional code in comparison with RFID is that it is read-only, while some RFID tags are read-write type.

(3) Small Area Size
    Comparing to barcode, the area size of 2-dimensional code can be small so that we can print a small lable and attach them to the backs of books. This is one of the two notable advantages of 2-dimensional code against barcode. They use the barcode labels for books in the National Library and other libraries in Singapore [9]. Their book labels are excellently designed; a label consists of barcode part and classification data part. They put the label so that the classification data part comes on the back part of the book and the barcode part comes on the cover part. With 2-dimensional code we can put these data on the back of the books with ease.

(4) Readable by and Displayable to Mobile Phones
    Mobile phones are excellent input/output equipment for 2-dimensional code. Many mobile phones are equipped with cameras and they are able to read the data in 2-dimensional code; typically QR code, for example. Also, in the other way, mobile phones can be used for displaying 2-dimensional code, which is used as paperless ticket, library's user ID, and others.

(5) Error Correction Facility
    2-dimensional code is designed so that even if part of the encoded area gets dirt or something and get impossible to read, the total information can be recognized with error correcting algorithm. This is a big advantage to barcode.

Fig. 5. Barcode (left) and 2-Dimensional Code: QR Code (right)

Conclusively 2-dimensional code system has advantages to as a recording media in terms of data size, recognision speed, area size, and others. It has advantages to RFID in terms of cost and readability by mobile phones. On the other hand RFID has advantages to barcode and 2-dimensional code in terms of locationing, ability to recognize the tags in hidden places, recognition speed, etc.

One of the good uses of advantage of 2-dimensional code in library is to put a label with 2-dimensional code to the back of a book so that the book's ID, its catalog data, and other related data can be recognized as the book is set on a bookshelf. Then such data are easily recognized with patrons' mobile phones and with code readers by the library staff for inventory and for other purposes.

3 Library Marketing System

The aim of the library marketing system is to help libraries with improving patrons' convenience and library management based on the data collected and stored in the libraries and with analyzing them with data mining methods.

In this section we consider two types of AIDC (Automatic Identification and Data Capture) technology, RFID and 2-dimensional code, as basic technologies that support library marketing.

3.1 Library Marketing System with RFID

Figure 6 illustrates a model for library marketing system with RFID, which is an extention to the system in Figure 1. In addition to the data that can be collected with barcode system, it can collect usage data for shelved materials by using intelligent bookshelf (IBS). Applicability to bookshelves is one of the most important advantages of RFID system because it is impossible to realize IBS with other types of AIDC technologies such as barcode, 2-dimensional code, etc.

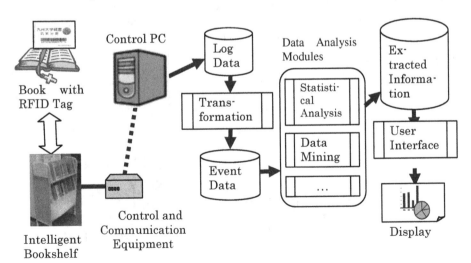

Fig. 6. An Experimental Library Marketing System with RFID

We have tried an experiment with the IBS in Figures 4 and 6 [13]. Original log data obtained from the IBS look as is shown in Figure 7. The first few lines are the headers from the controller system, followed with command lines. A command line start with the timestamp data followed with the command and its responding data, which are quoted with double quotation marks.

The command part starts with "<" character and the respond part with "<" character. The string "ST" stands for the status of the bookshelf. The following data proves status, shelf ID, reading time, the number of detected tags, and the list of tag IDs with 16 characters. For example, the first long line indicates that 9 tag IDs are detected and the IDs are "E00401000314CBA2," "E00401000314E148," followed by several other IDs and lastly "E004010003150040."

The log data are redundant because the status will not change for a long time we consider the frequency of data sampling and thus the detected tag IDs are same in a long line of data. So it is better to transform this log data format into the format that can be dealt with ease. We call the data in this transformed format the event data. The transformation program converts the original log data into event data.

The event data can be represented as a list of events as is shown in Figure 8; the first two columns are timestamp data that consist of date and hour-minute-second, followed with two columns of shelf ID, i.e. tier ID and shelf ID, one column of tag ID, and the type of event, or the change of status, which is either IN or OUT. The status IN means that the book is put on the denoted shelf and OUY means the other way.

```
*,2006/09/25 15:55:47, 0 ,"Log Start"
*,2006/09/25 15:55:47, 0 ,"TCP/IP Connection"
*,2006/09/25 15:55:47, 0 ,"Local Port:0, Remote Host [172.18.25.112:10001]"
<,2006/09/25 15:55:48, 6 ,">ST00"
>,2006/09/25 15:55:56, 20 ,"<ST0011000005500000"
>,2006/09/25 15:55:58, 20 ,"<ST0012000005590000"
>,2006/09/25 15:55:59, 20 ,"<ST0013000005500000"
>,2006/09/25 15:56:01, 164 ,"<ST0014000005E20009E00401000314CBA2E00401000314E1
48E004010003150DB9E00401000314F6ADE004010003150DBBE00401000314D735E00401
000314EA2DE00401000315382FE004010003150040"
>,2006/09/25 15:56:02, 20 ,"<ST0015000005500000"
>,2006/09/25 15:56:04, 20 ,"<ST0011000005500000"
>,2006/09/25 15:56:05, 20 ,"<ST0012000005590000"
>,2006/09/25 15:56:07, 20 ,"<ST0013000005500000"
>,2006/09/25 15:56:08, 164 ,"<ST0014000005B10009E00401000314CBA2E00401000314E1
48E004010003150DB9E00401000314F6ADE004010003150DBBE00401000314D735E00401
000314EA2DE00401000315382FE004010003150040"
```

Fig. 7. A Sample Original Data of the Intelligent Bookshelf Used in the Experiment

To formulate, the i-th event data collected in this IBS is represented as a quadruple in the following format:

$d_i = (In/Out, Shelf-ID, Tag-ID, Time)$

where "In/Out" specifies book is stored in a bookshelf or removed from a bookshelf, "Shelf-ID" and "Tag-ID" specify which shelf the book is stored to or removed from and which book is it, respectively. The "Time" is the timestamp which specifies date and time of this status change.

2006/6/2	15:57:06	1	2	E004010003153ED1	IN
2006/6/2	15:57:20	1	2	E004010003153ED1	OUT
2006/6/2	16:06:38	1	2	E00401000314EA68	IN
2006/6/2	16:06:38	1	2	E00401000314D6F0	IN
2006/6/2	16:06:53	1	2	E00401000314EA68	OUT
2006/6/2	16:07:15	1	2	E00401000314EA68	IN
2006/6/2	16:07:29	1	2	E00401000314EA68	OUT

Fig. 8. A Sample Event Data

The system log file is a list of, or a collection of, such data. Initially it is sorted according to Time. If we extract the data by specifying the Tag-ID, say TID, we get the following collection of data.

$D(TID) = \{ (In/Out, Shelf\text{-}ID, Time) \mid d_i = (In/Out, Shelf\text{-}ID, TID, Time) $ *for some In/Out, Shelf-ID, and Time}*

From this data, we can get the frequency of the usage about book which has the specified tag-id "TID". We can also get information something like, which time zone of a day, which day of the week it is read most often, etc.

The most important part of the library marketing system is the data analysis, or assisting data analysis, part because the usefulness of the system depends mostly on the capability of extracting useful information and/or knowledge for library patrons and library staff. However we have to point out here, that the extracted information/knowledge is basically a guess. We have to start with a low level guess and to improve its accuracy.

Among the analysis methods we take some possibly useful information that is extracted from the usage data.

(1) Frequencies of Use

Suppose we have a list of event data that are collected in a week. For a fixed material (i.e. book, magazine, CD, etc.) an OUT-IN pair of data for it indicates that it is taken out from a shelf and is returned afterward. This is a session of use of the material.

By counting the number of sessions for this material we can get how many times the material is used in this week. This information is helpful for both patrons and library staff. A patron can use this information when she or he wants to find a material that would be helpful for her or his study. We can suppose that a material that is used, i.e. probably means is read, by many people will be more useful than other ones.

The library staff can use this information when they choose which books to purchase. By analyzing the frequency data together with their classification number and other data, librarians can get information what kind of books are more used than other kinds. If they can guess by using this information that a specific book might be used very much in advance they can purchase more than one volumes of it.

Furthermore a material which is used very much will be damaged faster than those which are used not so much. So by accumulating the frequencies of use of a material, librarians can use the data and decides when to check it for restoration and when to discard.

(2) Accumulated Time of Use

Suppose again that we have a list of event data for a week. This time, for each material, we sum up all the session times, where session time is the duration of time from OUT-time to IN-time of the material.

A patron can use this information as an index for usefulness of material. If a material has long duration time, it might mean that it is more useful for students. Of course it is possible that a book which has just a short duration time even though it is potentially highly useful because of that it is just purchased and/or it is now well known to many students. However the duration time and usefulness should have some correlation each other.

Frequency and duration time are similar in a sense because both of them reflect popularity in some senses. The big difference lies that the former is an index for attractiveness of the material. If a material looks good in a first glance, it will be taken out very often and thus the frequency is high. However if it is not a useful book for the students, it will be returned in a short time and the total duration time will not be very long.

(3) Max, Min and Average Session Times

The maximum, the minimum and the average session times of a material will be able to be used to know how the material is used by students; very roughly. If the average time is very close to the maximum time of the sessions of the material and it is a long time, it means that the material is used for a long time, most of time it is used. Then we can guess the material is used as a textbook or something like that.

On the other hand, the average time is close to the minimum time and it is relatively short, then it may mean that the material is normally used like a dictionary or something like a reference book.

In order to estimate more precisely and accurately we will need more information about the material. However just by analyzing the usage data collected by an IBS, we can extract such estimated information and it will be helpful for patrons, or students, when they need to decide which material to use and in what way they use it.

Such information extracted from usage data is also helpful for library staff when they decide which material to purchase, when they decide where and in what way they put the materials. By combining this data and information with other information such as circulation data, catalog data of materials, we will be able to extract more information and extracted information and knowledge will become more helpful not only for students but also for librarians. We have to keep researching on this issue as one of the most important topics.

3.2 Library Marketing System with 2-Dimensional Code

As has been pointed out, quite a lot of mobile phones have equipped with cameras and can recognize 2-dimensional code such as QR Code. This is a great advantage for 2-dimensional code in comparison with RFID based library marketing system.

Let us suppose we put 2-dimensional code label to books in a library. The shelves might look like the one shown in the left part of Figure 9. A patron can identify the ID of a book that he or she is interested in. By aiming the mobile phone and take a picture of the 2-dimensional code, the book's ID data is easily taken into the mobile phone. Then the ID data is transmitted to the server for it, and the information relating to the book will be displayed on the phone. Starting from this point, the patron can go further. For example, he or she can get further information, put the data to his or her own blog, record the book to the personal virtual shelf, and even go to an e-commerce site and purchase the book.

One possible URL format recorded in the 2-dimensional code for such service is as follows:

http://<SeverIP>/bookinfo.cgi?id=<BookID>

It should be good if the system can recognize the reader's ID in order to differenti-ate the responses to the terminals. One possible mechanism is the use of security QR code [11], with which a part of a QR code data can be recorded as hidden data and can be recognized by some designated terminals only. In our case, the information that only concerns with the library jobs are recorded as such a hidden data so that the terminals for library jobs only can use such information.

Another possible mechanism is to organize the system so that when a terminal connects to the server, it sends its ID to the server as well as the ID of the material. The URL format for this might be like this:

http://<ServerIP>/bookinfo.cgi?bookid=<BookID>,patronid=<PatronID>

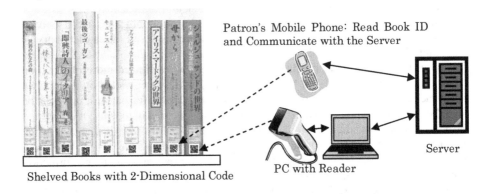

Patron's Mobile Phone: Read Book ID and Communicate with the Server

Server

Shelved Books with 2-Dimensional Code

PC with Reader

Fig. 9. Books with 2-Dimensional Code Label Together with Data Collection and Service

Whatever methods we choose, these actions will be recorded in the database of the server and will be used as a basic data for library marketing.

The combination of the PC and the reader in Figure 9 indicates the jobs by library staff. It will be used for inventory, search for misplaced books, and maybe more. Be-cause of the very high cost for RFID system, some library may choose 2-dimensional code system as a substitute for the barcode system which they are using now. In such a case PC with 2-dimensional code is a good choice for such library works.

3.3 Library Marketing System with Combined Data

In the previous section we discussed from the viewpoint that 2-dimensional code is an alternative to RFID because it is an upgraded version of barcode and is cheaper than RFID.

In this section we would like to discuss from the viewpoint that it is a complementary system to RFID. The most important advantage of 2-dimensional code is its affinity for mobile phones. Here again, we would like to emphasize the advantages of mobile phones for our purpose such as:

(1)   Quite a lot of mobile phones are quipped with cameras and are able to recognize the data printed in the form of 2-dimensional code attached, for example, on the books. Also almost all patrons have mobile phones these days and carry them all the time. Furthermore they have buttons for input and screens for output. Thus mobile phones are universal I/O devices for information, or PDAs, that support our ubiquitous life. This is the reason that 2-dimensional code is very convenient for recording and providing various kinds of information.

(2)   Mobile phones are communication devices in nature. They can transmit and receive data to and from the servers in the Internet. Thus the essential information printed on the 2-dimensional code labels is their IDs, or keys, to the other information. We can get all information based on such key data.

(3)   Each mobile phone will be used basically one person; i.e. its owner. So it is a good tool for individual authentication, thus it is very useful as a tool for providing personalized library services to patrons.

As we consider the current status of arts, RFID based IBS is the most appropriate tool for recognizing the book IDs and 2-dimensional code is the one for recognizing the patron IDs. By applying these two types of AIDC technologies the library marketing system can collect data about patrons' interest to the shelved books and the books that attract interest of each patron.

One possible patron service in this situation is a personalized recommendation of reading materials to a patron. First, the system recognizes the patron's interest type, or profile, by analyzing the interest data through 2-dimensional code label and from other methods. Then the system finds the books that are attractive to many people and will match to the patron's profile on interest. The attractiveness can be evaluated based on the access frequency data from the IBSs, from the circulation data, and from other sources. The matching criterion may vary. One possible one is to check the similarity of keywords based on the meaning and the classified field relating to the keywords.

Another combination candidate is with patrons' entrance and exit data of the target library. Even if the two frequency numbers of access are the same, the one should be considered more important than the other if the number of patrons staying in the library is smaller than the other. So the number of patrons staying in the library can be used for standardizing the frequency data from RFID and 2-dimensional code.

4 Concluding Remarks

The aim of this paper is to present two types of library marketing systems; with RFID and with 2-dimensional code. Firstly we describe the basic concept of library marketing, which consists of two important parts; data collection and data analysis, or data mining. RFID is an excellent tool for data collection. By installing the RFID readers to bookshelves, we can construct intelligent bookshelves (IBSs), with which we can collect usage data of library materials automatically. We presented three types of analysis methods for such usage data.

2-dimensional code has characteristics inbetween RFID and barcode in terms of data collection technology. The most important advantage of this technology is that quite a lot of mobile phones are equipped with cameras so that they can be used as readers for 2-dimensional code. Also mobile phones have screen and are able to communicate with application servers, thus they can be used as output devices for library services. For utilizing this advantage, we propose an idea of putting labels with 2-dimensional code to books so that patrons can get information about the books they are attached on.

One of the best candidates for library marketing may be a combination of these two types of technologies. The library system can collect usage data of library materials with RFID-based intelligent bookshelves and the patrons can get services that use the extracted information, knowledge, and know hows from these data.

Now we are preparing to realize a library marketing system based on the concept presented in this paper. Some of my students have started developing a system for 2-dimentional code. In collaboration with some librarians in Japan and Korea, we are planning to make an experiment for collecting usage data with intelligent bookshelves.

Our future plan includes:

(1) to develope a prototype library marketing system by combining RFID and 2-dimensional code technologies,
(2) to do experiment for automatic data collection,
(3) to develope useful analysis methods, and
(4) to prove the usefulness of the concept of library marketing so that it is installed in some libraries.

References

1. AIM Global: http://www.aimglobal.org/
2. City of Kita Library (in Japanese):
   http://www.library.city.kita.tokyo.jp/
3. Finkenzeller, K.: RFID Handbook, 2nd edn. John Wiley & Sons, Chichester (2003)
4. Kim, E.J.: A Study on the Space Organization by the User's Behavior in Public Library, Doctoral Theis, Gyeonggi University (in Korean) (2009)
5. Kyushu University Library: http://www.lib.kyushu-u.ac.jp/
6. Minami, T.: RFID Tag based Library Marketing for Improving Patron Services. In: Hoffmann, A., Kang, B.-h., Richards, D., Tsumoto, S. (eds.) PKAW 2006. LNCS (LNAI), vol. 4303, pp. 51–63. Springer, Heidelberg (2006)

 7. Minami, T.: On-the-site Library Marketing for Patron Oriented Services. Bulletin of Kyushu Institute of Information Sciences 8(1), 15–33 (2006) (in Japnese)
 8. Minami, T.: A Perspective to the Library in Network-Oriented Society – Ubiquitous Library Services through PDAs. Bulletin of Kyushu Institute of Information Sciences 10(1), 1–17 (2008) (in Japanese)
 9. National Library Board Singapore: http://www.nlb.gov.sg/
10. Ranganathan, S.R.: The Five Laws of Library Science, Bombay. Asia Publishing House (1963)
11. Security QR Code (in Japanese): http://www.denso-wave.com/ja/adcd/product/qrcode/sqrc/index.html
12. University of Nevada, Las Vegas Libraries: http://www.library.unlv.edu/
13. Zhang, L., Minami, T.: Library Marketing that Boosts Education with RFID. In: Proc. 6th International Conference on Information Technology Based Higher Education and Training, ITHET 2007 (2007)

Knowledge Audit on Special Children Communities

Aida Suzana Sukiam, Azizah Abdul Rahman, and Wardah Zainal Abidin

Department of Information Systems,
Faculty of Computer Science and Information Systems,
Universiti Teknologi Malaysia, 81310 Skudai Johor
{aidasuzana,azizahar,wardah}@utm.my

Abstract. This paper reports on how knowledge audit analysis was conducted for special children (SC) communities in Malaysia context. The purpose of the knowledge audit is to determine the knowledge requirement of SC communities which may reveal the required, available and missing knowledge and the person involved related to SC. Four communities of practice (CoPs) have been identified namely Parents, Educators, Medical Experts and Researchers. The knowledge audit has been conducted in four phases. There are Knowledge Needs Analysis, Knowledge Inventory Analysis, Knowledge Flows Analysis and Knowledge Mapping. Questionnaires and interviews were conducted as knowledge audit tools for facilitating the collection of data, information and evident. In order to represent the knowledge audit results, matrices presentation is used. These representations help in tracing knowledge and verify the results with the CoPs more clearly. The result shows that knowledge audit yields a number of benefits that include the missing gap between required and available knowledge. The requirement can be used to develop a one stop center for SC communities to communicate among other CoPs. Subsequently, recommendations can be derived for better managing of the knowledge.

Keywords: Knowledge Audit, Communities of Practice and Special Children.

1 Introduction

Special Children with learning disabilities are different from typical children. They are unable to fend for themselves in doing any decision making due to their conditions, without the help from parents, teachers or others. Referring to Persons with Disabilities *Act* 2007, disabled person is identified as *"those who have long term weaknesses in phisical, mental, intelectual or sense which prevent them from full and effectifely participate in the society "*[1].

In order to ensure the children get the best nursing and education based on their condition, there are many requirements of relevance knowledge that SC community needs. However, in Malaysia context, the provision of this knowledge in documented form such as books, journals, websites or others are very limited. Most of the knowledge is remain in the CoPs' mind as they gained that knowledge based on experiences. Different group of people in SC community may hold different type of knowledge and carry out different types of processes. Therefore, they need to acquire knowledge from others in order to perform their daily work especially the parents.

D. Richards and B.-H. Kang (Eds.): PKAW 2008, LNAI 5465, pp. 198–207, 2009.

There are so many questions raise up. They need the support and help from people who understand their situation.

Thus, it is important for SC communities to interact and communicate with each other to share their knowledge and experience. Therefore, we need to identify what is the relevant knowledge acquired by the SC community. It is the intention of this paper to determine the required, available and missing knowledge, the person involved and the government agencies related to SC. For this purpose, the authors have been conducted the knowledge audit analysis. A result from the knowledge audit analysis is reported in this paper.

2 Literature Review

This section explains the fundamental concepts of community of practice and knowledge audit.

2.1 Community of Practice

In order to practice knowledge sharing, community of practice offers a way to theorize tacit knowledge which cannot easily be captured, codified and stored [3]. Communities of practice are groups of people who share a passion for something that they know how to do, and who interact regularly in order to learn how to do it better [4]. This definition is inline with Debowski's definition towards CoP, where she defined the CoPs as groups of people with common interests who meet to share their insights in order to develop better solutions to problems or challenges [5]. According to Wenger, there are three fundamental characteristics of communities. The characteristics are:

- **Domain**: the area of knowledge that brings the community together, gives it its identity, and defines the key issues that members need to address.
- **Community**: the group of people for whom the domain is relevant, the quality of the relationships among members, and the definition of the boundary between the inside and the outside
- **Practice**: the body of knowledge, methods, tools, stories, cases, documents, which members share and develop together.

2.2 Knowledge Audit

Liebowitz defines a knowledge audit as a tool that assets potential stores of knowledge [6]. It is the first [6] and critical part of a knowledge management methodology [7]. The knowledge audit examines knowledge sources and use: how and why knowledge is acquired, accessed, disseminated, shared and used [8]. The main purpose of knowledge audit is to determine what it knows, what it does not know, what it needs to know and how it should go about improving the management of its existing knowledge [9].

For this study, the knowledge audit have been conducted to determine the knowledge requirement of SC communities which may reveal the required, available and missing knowledge, the person involved and the government agencies related to SC. There is no universally accepted approach to K-Audit [7]. Based on multiple groups of practitioners' experience [6, 10-12], the authors adapted the knowledge audit tools and method that has been proposed and tested by those practitioners. Both tools and method were subject to iterative refinement suggested by practitioners during the field trial. As proposed by them, knowledge audit should be divided into components or activities which are ideally performed in sequence.

3 Methodology

As mentioned in previous section, knowledge audit can be divided into components which result in milestone for the purpose of diagnostics and corrective measures. The authors adapted the knowledge audit method proposed by several practitioners [6, 10-12] as depicted in Fig. 1.

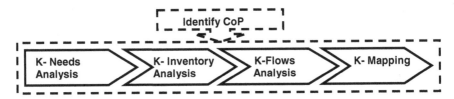

Fig. 1. Principal Component of Knowledge Audit

There are four components namely: Knowledge Needs Analysis, Knowledge Inventory Analysis, Knowledge Flows Analysis and Knowledge Mapping. These activities will be performed after the authors identify the CoP related to SC. A knowledge audit needs to be performed for each CoP and not necessary need to be conducted in sequence.

3.1 Knowledge Needs Analysis (K-Needs Analysis)

The needs analysis is a process by which information users are asked on what information resources or services that they needed to perform their job [13]. The major goal of this task is to identify precisely what knowledge the CoP possess currently

Table 1. K-Needs Analysis Activities

Activities	Objectives	Instruments/Tools	Deliverables
Background Study	- To understand the CoP/related agencies to be audited. - Define CoP/related agencies' goal & objectives.	- Questionnaires - Interviews - Related Portal	- List of goals & objectives - Background information related to the audit such as organizational chart & workflow
Draft an initial K-Needs Matrix	Identify precisely what knowledge CoP/related agencies:	- Pilot surveys	- Initial K-Needs Matrix
Verify K-Needs Matrix	- Currently posses - Currently/Future required - Currently/willing to share	- Surveys - Interviews	- Final K-Needs Matrix.
Reporting	To report K-Needs Analysis	- Verified K-Needs Matrix	- Knowledge Asset Document

and what knowledge they would require from others in order to meet their objectives and goals[12]. It reveals knowledge that they are willing to share with others. Table 1 shows the activities involved in this task.

3.2 Knowledge Inventory Analysis (K-Inventory Analysis)

Knowledge inventory is knowledge stock-taking to identify and locate knowledge assets and resources throughout the entire organization. Knowledge inventory comprises of two entities: Physical (Explicit) Knowledge Inventory and Experts (Tacit) Knowledge Inventory. In this case, the experts refer to the educators of the SC. The K-Inventory analysis involved three major activities as shown in Table 2.

Table 2. K-Inventory Analysis Activities

Activities	Objectives	Instruments/Tools	Deliverables
Identify related portal, database and document.	To identify possible knowledge location	- Pilot Survey - Related portal - Interview	- Listing of the knowledge location
Draft an initial K-Inventory Matrix	To identify & locate k-assets & resources.	- Pilot surveys	- Initial K-Inventory Matrix
Verify K-Inventory Matrix		- Surveys - Interviews	- Final K-Inventory Matrix.
Reporting	To report K-Inventory Analysis	- Verified K-Inventory Matrix	- Knowledge Asset Document

3.3 Knowledge Flows Analysis (K-Flow Analysis)

Knowledge flow analysis focus at how knowledge resources move around the organization, from where it is to where it is needed [12]. According to Sharma and Chowdhury, by perform this analysis; it may reveal how CoP find the knowledge they need and how they share the knowledge they have and some barriers to effective flows. Such analysis looks at people, processes and systems. For this paper, the authors look into the processes which to examine how CoP go about performing their daily work activities and how knowledge seeking and sharing form parts of those activities. Table 3 shows activities involved in order to perform this analysis.

Table 3. K-Flows Analyssis Activities

Activities	Objectives	Instruments/Tools	Deliverables
Draft an initial K-Flows Matrix	To determine: - how CoP/related agencies find needed knowledge. - how they share the knowledge they have.	- Pilot surveys	Initial K-Flows Matrix
Verify K-Flows Matrix		- Surveys - Interviews	Final K-Flows Matrix.
Reporting	To report K-Flows Analysis	- Verified K-Flows Matrix	Knowledge Asset Document

3.4 Knowledge Mapping (K-Mapping)

Knowledge map is used in order to visualize CoP/related agencies knowledge based on the K-Needs, K-Inventory and K-Flows Matrices report. It demonstrates who has knowledge, where these persons are located and with who they most often share and exchange knowledge[11]. Table 4 shows the activities involved in order to perform knowledge mapping.

Table 4. K-Mapping Activities

Activities	Objectives	Instruments/Tools	Deliverables
Study K-Needs/K-Inventory/K-Flows Matrices Report	- To locate important knowledge - To show knowledge user where to find the knowledge in visual mode.	- K-Needs/K-Inventory/K-Flows Matrices Report	- Analysis of K-Needs/K-Inventory/K-Flows Matrices Report
Draft an initial K-Map		- Suitable K-Map Tools	- Initial K-Map
Verify K-Map		- Focus Group	- Verified K-Map
Reporting	To report K-Map	- Verified K-Map	- Knowledge Asset Document

4 Results

This section reports on the results of knowledge audit analysis in SC educator's context. For a start, the related CoP was identified. Subsequently, the knowledge audit was conducted for each CoP.

4.1 Identification of Communities of Practice

Four major CoPs have been identified in this context. There are Parents, Educators, Medical Experts and Researchers in area of LD as shown in Fig.2. However, this paper reports only on Educators context.

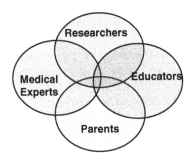

Fig. 2. Communities of Practice

4.2 Knowledge Audit Results

A number of knowledge audit tools have been used in order to determine SC communities needs (K-Needs Analysis). Up to this stage, knowledge audit have been conducted to parents and educators. To have better understanding on what are the processes involve, some literature on related portal have been reviewed and a few sets of questionnaires were distributed to the possible respondents based on the identified CoPs. Table 5 shows the focus of the questions asked in the questionnaires and the purpose of asking them.

Table 5. The Purpose of Questions Asked

Questions	Purpose
List the processes that involve SC community?	To have a better understand on what are the processes
Describe the process flows (for each process)	involve each CoP pertaining to SC.
List the future, currently required, available & shared knowledge to perform those processes?	- To identify all processes performed by CoP - To identify knowledge needed to perform each process.
What is the format of this knowledge?	To identify tacit and explicit knowledge.
From whom you acquire those knowledge?	This information is for constructing knowledge flow,
Who are the users for this knowledge?	inventory and map
What are the mechanisms that you used to acquire this knowledge?	To identify communication media used by the CoP. This information can be used to identify mode of
Knowledge sharing mechanisms currently use?	interactions between the CoP and type of knowledge
Preferred mechanisms to acquire knowledge?	To determine the most suitable KS mechanism (based
Preferred mechanism to share knowledge?	on LR and preferred mechanism by the CoP)

To present the findings in a systematic way, matrices presentation were used. These matrices also used for verification purpose. This process was subject to iterative refinement in order to produce the final matrices. Table 6 shows partial results from the K-Needs analysis matrix in SC Educator's context. This matrix yields that there is a number of required knowledge needed by educators in order to perform their job.

Table 6. K-Needs Analysis Matrix for SC Educators

| Process | Knowledge | | |
	Current Exist	Current Required	Future Required
Diagnostic Assessment	-Collaboration information between teacher & related department. -Assessment process -Specialist Advice	Complete information on characteristic of SC registered in the school.	Standard guideline on assessment and measurement level of children with LD
Registration (Education Department)	Process on placement of SC in special class suit with their qualification	Assessment process did by medical experts before SC will be placed in certain school	Detail information on SC background (from birth to date) before they are placed in certain school
Placement of SC in special class	Guideline for placement, Behavioral, Developmental and Achievements Test		
Teaching and Learning (T&L)	Syllabus for every level of LD which need to teach and finish for the whole year	Syllabus of subject for high level and skillful children with LD	UPU level

There are several activities involved in each process as listed in K-Needs matrix (see Fig 3). Three activities are involved to perform Diagnostic Assessment process. There are screening, diagnosis and assessment process. The research team was numbered each activity and refers it as a *Task No*. These tasks were revealed from the flowcharts which were gathered during the background study.

Fig. 3. Diagnostic Assessment Flowchart

In order to identify whether the knowledge is already exist and where the knowledge is located, the K-Inventory Analysis Matrix need to be formulated. Since this research was conducted to the society and not to the organization, the locations of the knowledge are referred to the owner of the knowledge as shown in Table 7. This table shows the K-Inventory Analysis Matrix for Diagnostic Assessment process.

Table 7. K-Inventory Analysis Matrix for SC Educators

PROCESS: Diagnostic Assessment					
TN	**Purpose/ Decision Making**	**Knowledge Item**	**Knowledge Owner**	**E/T**	**Document needed (if explicit)**
1	To consider if the child needs further assessment	Child's behavior and skill	Parents	E	Special Needs Children Registration & Placement Suggestion Form (0-18 year old)
			SET		
			SED's representatives		
			Medical Authorities		
2	-To assess the child. -To provide a diagnosis	Child's behavior and skill	Medical Authoroties	E	M-Chat, Denver Schedule
			SET		
		Symptom	Medical Authoroties	E	
			SET		
3	-To assess the development and academic needs of the child - To get confirmation by medical authorities - To document child's behavior & abilities - To write IEP	Child's behavior and skill	Medical Authorities	E	Individual Education Plan (IEP)
			SET		
		Symptom	Medical Authorities	E	
			SET		
		Academic needs of the child	Medical Authorities	E	
			SET, SED		
		Education plan	Medical Authorities	E	
			SET, SED		
TN- Task No; **E**-Explicit; **T**-Tacit; **SED**- Special Education Department; **SET**-Special Education Teacher					

By making comparison between earlier analysis of K-Needs and K-Inventory Matrix, the authors were identified the missing knowledge required by the educators. For example, the educators still required the complete information on SC's characteristics at their school. However, this knowledge already exists and owned by the medical authorities.

Once the location of the knowledge already identified, the authors need to determine how the knowledge moves around from where it is to where it is needed. Therefore, the K-Flows Analysis Matrix was used (See Table 8). This matrix is used to determine how people in an organization find the knowledge they need, and how do they share the knowledge they have.

Table 8. K-Flows Analysis Matrix | Education Context

TN	Knowledge item	Knowledge Owner	Knowledge User	Communication Media
	PROCESS: Diagnostic Assessment			
1	Child's behavior and skill	Parents	Parents, SED,SET, Medical Authorities	Face-to-face, Phone, Formal letter
		SET, SED's representatives		
		Medical Authorities		
2	Symptom (Diagnosis)	SET	SET, Medical Authorities	Face-to-face
		Medical Authorities		
			Parents	Face-to-face
3	Child's behavior and skill	SET	SET, Medical Authoroties	Face-to-face
		Medical Authorities		
			Parents	
	Symptom (Diagnosis)	Medical Authorities	SET, Medical Authoroties	Face-to-face
			Parents	
	Academic needs of the child	SET	Medical Authorities	
		Medical Authorities	SET	Formal letter
			Parent	Face-to-face
	Education plan	Special Education Teacher	Medical Authorities	
		Medical Authorities	SET	Formal letter
			Parent	Face-to-face
SED- Special Education Department; **SET**-Special Education Teacher				

Based on the interpretation and understanding of K-Needs, Inventory and Flows Analysis Matrix, the authors were built the knowledge map as depicted in Fig.5. This map is used in order to visually represent the knowledge. In order to perform diagnostic process, each CoP communicates among the others to share the knowledge they have.

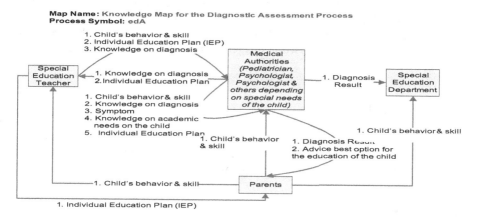

Fig. 4. Knowledge Mapping | Education Context

Box at the tail of the arrow shows the owner of the knowledge, and box at the head of the arrows shows who needs that knowledge. For example, Special Education Teacher owned three types of knowledge that needed by the Medical Authorities. The knowledge is (i) Childs behavior and skills; (ii) Individual Education Plan and (iii) Knowledge on diagnosis. Vice versa, the Medical Authorities also owned some knowledge that needed by the Special Education Teacher. There are (i) Knowledge on diagnosis and (ii) Individual Education Plan.

5 Discussion and Conclusion

Based on the researches that were done, the authors revealed that SC knowledge is very complicated due to involvement from many CoPs. Different group of people in SC community may hold different type of knowledge and carry out different types of processes. Therefore, there is a need to conduct knowledge audit in order to reveal those knowledge. Knowledge audit allow investigating those processes and knowledge involved in each process more details.

The paper has reported on knowledge audit result in educator's context conducted by the authors. The approach for knowledge audit presented in this paper consists of four components proposed and practiced by the other practitioners. The components are K-Needs Analysis, K-Inventory Analysis, K-Flows Analysis and K-Mapping. In order to perform these components, sets of questionnaires were distributed to the participants and followed by the interviews session.

The results show that the knowledge audit yields a number of potential benefits including the identification of the available, required and missing knowledge and the subsequent recommendations of KM strategy that can be used for better managing the knowledge. Furthermore, the benefits will impact the people in SC community in different ways. For future work, the requirement can be used to develop a one stop center for SC communities to communicate among other CoP.

The knowledge audit results were represented in matrices presentation. The advantage of using these matrices is easy in tracing knowledge and verifies the results with the CoPs more clearly. Due to the iterative process, the missing knowledge in each matrix can be reduced while verification stage. As a conclusion, the authors revealed that, knowledge audit method used in this paper are systematic and suitable to the SC community's environment.

Acknowledgement

The research team would like to express their sincerely thank to eScienceFund (Project No: 01-01-06-SF0221), Ministry of Science and Technology Innovation (MOSTI) for the financial support of the research work.

References

1. SWD., Rang Undang-Undang Orang Kurang Upaya, S.W. Department., Editor, p. 37 (2007)
2. Masyarakat, P.R.J.K.: (2008) (cited 2008), http://www.jkm.gov.my/
3. Wenger, E.: Knowledge Management as a doughnut: Shaping your knowledge strategy through communities of practice. Ivey Business Journal, 1–8 (2004)

4. Wikipedia, C.: Community of practice (2008) (cited November 2007),
   `http://en.wikipedia.org/w/`
   `index.php?title=Community_of_practice&oldid=190081904`
5. Debowski, S.: Knowledge Management. Wiley, Australia (2006)
6. Liebowitz, J., et al.: The Knowledge Audit. Knowledge and Process Management 7(1), 3–10 (2000)
7. Chong, D.Y.Y., Lee, W.B.: Re-Thinking Knowledge Audit: Its Values and Limitations in the Evaluation of Organizational and Cultural Asset. In: Knowledge Management in Asia Pacific Conference (2005)
8. Perez-Soltero, A., et al.: Knowledge Audit Methodology With Emphasis On Core Processes. In: European and Mediterranean Conference on Information Systems, pp. 1–10. EMCIS, Costa Blanca (2006)
9. Awad, E.M., Ghaziri, H.M.: Knowledge Management. An Integrated Approach. Prentice Hall, Englewood Cliffs (2004)
10. National Library for Health NHS. Knowledge Management Specialist Library-Auditing knowledge (2008) (cited),
    `http://www.library.nhs.uk/KnowledgeManagement/`
    `SearchResults.aspx?catID=10397`
11. Hylton, A.: A KM Initiative is Unlikely to Succeed Without a Knowledge Audit (2002) (cited)
12. Sharma, R., Chowdhury, N.: On the USe of a Diagnostic Tool for Knowledge Audit. Journal of Knowledge Management Practice 8(4) (2007)
13. Henczel, S.: The Information Audit as a First Step Towards Effective Knowledge Management: An Opportunity for the Special Librarian. In: Global 2000 Workdwide Conference on Special Librarianship, pp. 210–226. INSPEL, Brighton (2000)

A Dance Synthesis System Using Motion Capture Data

Kenichi Takahashi and Hiroaki Ueda

[1] Hiroshima City University, Department of Intelligent Systems,
731-3194, Hiroshima, Japan
{takahasi,ueda}@hiroshima-cu.ac.jp

Abstract. The aim of this system is to synthesize the choreography of the dance motions automatically by using the subjective music impression given by a user, even if the user has no knowledge to the dance at all. In order to show that the system is useful, we perform experiments. In the experiments, the dance data of 16 measures of two musics is synthesized, and the synthesized result of dancing data by the system and the results synthesized by a dance expert are compared.

Keywords: Motion capture, dance synthesis.

1 Introduction

The advent of the motion capture system has made us measure accurately human physical exercises and actions. As one of applications of motion capture, systems that digitally record human physical actions such as dance and represent the data on a computer by computer graphics have been studied so far. It seems that the real action expression which is used in recent computer games of these days reflects such evolution in information technologies[1]-[4]. EvaHiRES system that is an optical motion capture systems have been introduced in our university[5]. In this research, we design the system that synthesizes dance data utilizing the motion captured dance data obtained by the Eva HiRES system. Dance data are synthesized so as to be appropriate for music according to the inputted music impressions. Also, we discuss the utility of the dance synthesis system through experiments. The aim of this system is to synthesize the choreography of the dance motions automatically by using the subjective music impression given by a user, even if the user has no knowledge to the dance at all. In order to show that the system is useful, we perform experiments. In the experiments, the dance data of 16 measures of two musics is synthesized, and the synthesized result of dancing data by the system and the results synthesized by a dance expert are compared.

This paper is organized as follows: In section 2, we describe how to make the dance data file by using the motion capture system, and section describes the overview of the dance synthesis system. Moreover, in section 4, we show the experimental result performed for estimating the dance synthesis, and finally we give some conclusions and future tasks.

D. Richards and B.-H. Kang (Eds.): PKAW 2008, LNAI 5465, pp. 208–217, 2009.

2   The Motion Capture System

The motion capture system Eva HiRES system is consists of a personal computer to collect the motion data, a monitor, a near-infrared light electronic flash to reflect markers effectively, a controller, a high-speed camera, and a work station for control. Figure 1 shows an example of marker setting. Data tracking determines the correspondence of each marker to the part of the body and revises the discontinuity of data.

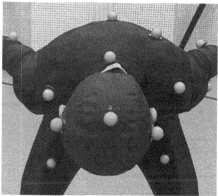

Fig. 1. An example of marker setting

Fig. 2. Examples of motion data bound with CG

Finally, we edit the track data in the temporal sequence of the three-dimensional coordinates of markers in order to make the data more complete. For the dance synthesis system, we collect the dance data for every two bars of music. The dance data is expressed by using CG software MAYA. In the experiments, the dance data of 16 measures of two music is synthesized, and the synthesized result of dancing data by the system and the results synthesized by a dance expert are compared. In order to bind the dance data captured by Eva HiRES system with a CG character, CG software Maya is employed. Figure 2 shows an example of the bind result. The obtained CG motion data is converted into the avi file that can be displayed on computers in Windows OS and for the dance synthesis system. Moreover, in order to make automatic synthesis of the dance data easier in the dance synthesis system, the adjustment of the number of frames and the unification of file formats are made with free software Aviutl.

3 The Dance Synthesis System

The objective of this system is to be able to synthesize automatically the choreography of the dance from the subjective impression given by a person that listens to music even if the person has no knowledge concerning the dance at all. In other words, the system chooses dancing data corresponding to the music impression and composes dancing motions just by inputting a subjective impression for music to this system. Figure 3 shows the flow of this system. In this research, music and the dance are first decomposed every two bars of music, the dance data of every two bars is

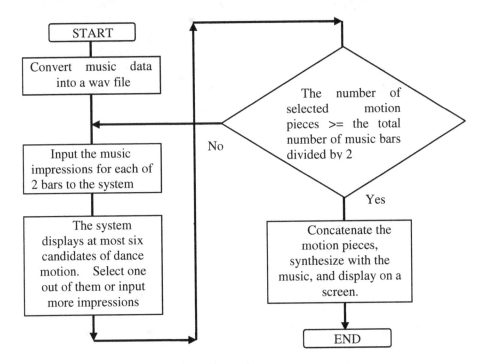

Fig. 3. The flow diagram of the dance synthesis system for the usage

collected by the motion capture system, and the data is expressed by CG. We attach subjective impression to each dance data(we call them an attribute of dance data.), and organizes it to a database. In this study, three subjective impressions (attributes) are attached to each of dance actions in the database, where we use 18 kinds of subjective impressions and 51 kinds of CG dance actions. Table 1 summarizes the 18 subjective impressions used in the experiment. In addition, a part of the dance data

Table 1. Summary of the 18 subjective impressions used in the experiment

ordinary	unique	small	big
simple	complicated	light	heavy
rambling	coordinated	soft	hard
weak	strong	strange	beautiful
dark	bright		

Table 2. Examples of a part of the dance data and their attributes

No.	Dance name	Attributes		
00	STEP1	big	simple	
01	SLIDE STEP	big	soft	
02	NEW JACK SWING	big	heavy	strong
03	CRAB	small	light	
04	SKEETER RABBIT	unique		
05	RUNNIG MAN	big	simple	unique
06	SQUARE	simple	soft	
07	SNAKE	beautiful	soft	
08	ENTRY	simple	light	
09	STOP & GO	coordinated	complicated	

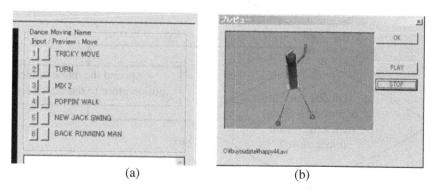

(a)   (b)

Fig. 4. (a) The list of candidates of dance motions, and (b) a preview screen of a checked dance data

and the attribute are listed in Table 2. Based on this database, the dance synthesis system composes the dancing choreographer corresponding to the subjective impressions for bars of music given by a user. Figure 4 shows the list and a preview screen of the extracted dance data. When a user put music impressions into the system through the interface, the system lists dance pieces corresponding to the impressions as in Fig.4. The user looks at dance motions through the preview and selects one. The algorithm for picking up dance pieces are shown in Fig.5.

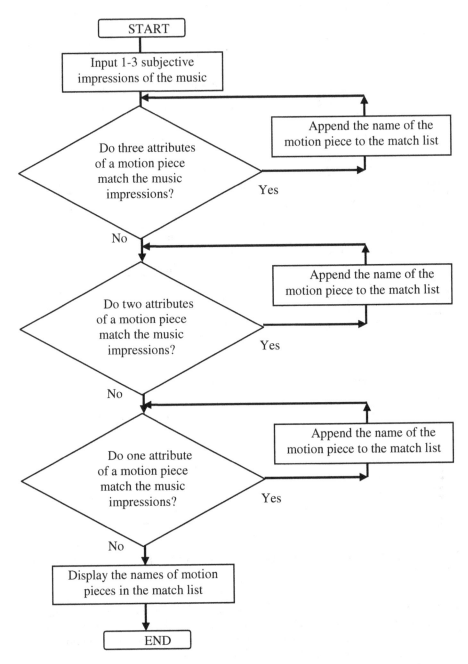

Fig. 5. A flow chart for selecting motion pieces that correspond to music impressions

4 Dance Synthesis Experiments

In order to show the utility of this system, a series of dance data was synthesized as a sample with the music, and they were displayed on a computer by Maya2.5. When this system is completed, we can obtain the choreography of the dance synthesized by the system just by inputting subjective impressions of music even if we don't have knowledge about the dancing at all. A composition file of a series of dance data is produced according to the following procedures.

(1) The dance data is extracted based on the subjective impression of music by using the dance synthesis system.
(2) For comparison, we also made a composition file in which the dance data is extracted by a dance expert.
(3) The dance data extracted in (1) is connected by using a software AviUtl[5]. Also, the dance data extracted by an expert is connected by using AviUtl.
(4) The music file of a wav format is read into the system.
(5) The system synthesizes the dance data and the music into a file, and store it in an avi format.
(6) The display speed is adjusted by using AviAdjst[6].
(7) The composition file is displayed by using Windows Media Player.

In the experiment, the dance data produced by the dance synthesis system and the dance data produced by an expert are compared and are estimated. The dance files are produced for two music, namely Zhigge and 2PAC. Each file includes eight dance data for 16 bars; One dance data corresponds to two bars of music. For the sake of the limited paper space, we summarizes the result for the first music. Tables 3 and 4 list the dance data in the first music Zhigge. Table 3 lists the dance data extracted by the dance synthesis system, and Table 4 lists the dance data extracted by an expert. In the table, shaded portions show that the attribute of a dance data by the system is the same as that by the expert, while the portions surrounded with thick lines show that the attribute of a dance data by the system is contradictory to that by the expert. The basic impressions of the music is as follows.

The impressions of Zhigge
It starts with the drum sound, and it becomes only the sounds of a synthesizer base instrument and the voice of the lap from the fourth bar; the impression is a "quiet" impression. However, from the 8th bar the intense drum sound that appeared till the 4th bar comes back, and lyrics of the lap and woman's singing voice are added.

Discussions
We can observe from Table 3 that both the dance data synthesized by the system and that by an expert extracted RUNNING and MAN(TWIST), while they appear in different bars of music. TURNNING 2STEP appears in the 16th bar in both the dance data by the system and that by an expert. Since basically Zhigge is a melody of an intense light atmosphere, a lot of dance data of step kinds is selected, and so both of dance data by the system and by an expert resemble closely. For instance, STEP1 is a step to kick backward, and 2STEP is a step to kick backward and forward. Considering from the subjective impressions of the music, it is reasonable that the dance

Table 3. The dance data extracted by the dance synthesis system for Zhigge

No. of bars	The system	Attributes		
02	NEW JACK SWING	big	heavy	strong
04	OVER ARM STEP	big	bright	simple
06	STEP1	big	simple	
08	BACK STREET1	simple	strange	
10	ARM'S MOVE	big	light	simple
12	RUNNING MAN(TWIST)	big	light	unique
14	ENTRY	simple	light	
16	TURNING 2STEP	big	coordinated	bright

Table 4. The dance data extracted by an expert for Zhigge

No. of bars	An expert	Attributes		
02	RUNNING MAN(TWIST)	big	light	unique
04	SKEETER RABBIT	unique		
06	WAVE2	dark	beautiful	
08	DOWN STEP1	heavy	beautiful	
10	JUMP STEP1	big	ordinary	strong
12	SIDE STEP	big	ordinary	
14	2STEP	bright	light	soft
16	TURNING 2STEP	big	coordinated	bright

synthesis system selected a lot of dance motion pieces whose attribute is "big"; Since this music gives in 1-4 measures and 9-16 measures an intense impression, the selection of a dance motion piece whose motion is big or large is reasonable. In 5-8 bars a lot of "simple" or "strange" dance pieces are selected, because the number of sounds is a few; STEP1 and BACK STEP1 at this time are monotonous choreographies that corresponds to the melody. On the other hand, the dance data extracted by an expert also includes dance motion pieces whose attribute is "big" or "unique" basically, and the impression of the choreography resembles that by the dance synthesis system.

We can observe from Table 3 that both the dance data synthesized by the system and that by an expert extracted RUNNING and MAN(TWIST), while they appear in different bars of music. TURNNING 2STEP appears in the 16th bar in both the dance data by the system and that by an expert. Since basically Zhigge is a melody of an

Table 5. The dance data extracted by the dance synthesis system for music 2PAC

No. of bars	The system	Attributes		
02	MIX2	coordinated	light	hard
04	TRICKEY MOVE	hard	strong	
06	POPPIN' WALK	hard	heavy	simple
08	DOWN STEP1	beautiful	heavy	
10	SLIDE	beautiful	light	coordinated
12	NEW JACK SWING	big	heavy	strong
14	BACK RUNNIG MAN	big	simple	strong
16	OVER ARM STEP	big	bright	simple

Table 6. The dance data extracted by an expert for music 2PAC

No. of bars	An expert	Attributes		
02	SMUFF	soft	heavy	
04	DOWN STEP1	beautiful	heavy	
06	POPPIN' WALK	hard	heavy	simple
08	AIR WALK	beautiful	light	soft
10	2STEP	bright	light	soft
12	BACK STEP2	simple	light	
14	BOUNCE(HOUSE)	bright	light	soft
16	MIX1	strong	bright	

intense light atmosphere, a lot of dance data of step kinds is selected, and so both of dance data by the system and by an expert resemble closely. For instance, STEP1 is a step to kick backward, and 2STEP is a step to kick backward and forward. Considering from the subjective impressions of the music, it is reasonable that the dance synthesis system selected a lot of dance motion pieces whose attribute is "big"; Since this music gives in 1-4 measures and 9-16 measures an intense impression, the selection of a dance motion piece whose motion is big or large is reasonable. In 5-8 bars a lot of "simple" or "strange" dance pieces are selected, because the number of sounds is a few; STEP1 and BACK STEP1 at this time are monotonous choreographies that corresponds to the melody. On the other hand, the dance data extracted by an expert also includes dance motion pieces whose attribute is "big" or "unique" basically, and the impression of the choreography resembles that by the dance synthesis system.

We also can see that the dance synthesis system makes choreography that resembles choreography produced by a human expert with his subjective impressions for the 2nd music 2PAC. The experimental result is listed in Tables 5 and 6.

This dance synthesis system extracts six dance motion pieces with attributes by using at most 3 music impressions, and a user selects one of them. Therefore, there is a possibility that the system extracts dance motion pieces to which inputted subjective impression are not necessarily corresponds, when the dance data with few matches with the inputted subjective impressions is chosen even if the user inputs three subjective impressions of the music. In the experiments, some dance motion pieces whose attributes are contradicting are selected. It occurs when the inputted subjective impressions are not reflected on selection. It is important that the impression received from the tune is not so different for people hearing the music while the choreography changes greatly depending on the dance choreographer. Through the comparative experiments, we see that there exists the subjective impressions suitable for the tune, a choreographer arranges dances according to his feeling, and the dance choreography is performed according to the subjective impressions. Therefore, the authors believe that the dance synthesis system can produce the choreography similar to the choreography produced by a human expert by employing the subjective impressions of the music.

5   Conclusions

We proposed a dance synthesis system that utilizes the motion capture data and synthesizes them with CG software Maya according to the inputted subjective impressions of the music. In order to show the utility of the system, we performed the composition experiments and compared the choreography produced by the system and that by an expert. In the experiments, 16 bars of dance data by the system and those by an expert are produced for two music. We showed the utility of the development of the dance synthesis system using subjective impressions common between the dance data and music.

Future tasks includes the revision of the dance data; each of the collected dance is two bars of length and can express only rough subjective impressions. The dance data is delimited to one bar or a more detailed unit, it is necessary to enable the selection of a lot of dance data. It may be necessary to divide the dance length from two bars to one bar and collect dancing data with a smaller unit and to make the system be able to choose more dancing data. After the collected dance data is delimited to one bar or a more detailed unit, the connection point of the dance pieces where a location where the connection may become unnatural depending on the combination part of data. In such a case, it is necessary to be able to compose dance pieces automatically so that at the connection point the move of a dance piece to another dance piece may be smooth and natural by extracting only dancing data that can be connected to the previous dance piece smoothly. Furthermore, the problem of the generality of the subjective impression is also enumerated. In this paper, we used the attributes collected and determined by an expert for 51 dance data. However, it is also useful to add a way that the attributes of the dance data can be given by users in case where the attributes are not suitable for the user feeling. Finally, in order to make automatic composition easier and smooth, it is desirable to develop a program to covert the speed of the dance motion to synchronize the music.

Acknowledgments. The authors would like to thank Mr.Kamimae and Mr. Nakamura for their help for motion capturing and valuable discussions on the system and experiments.

References

1. Okamoto, H., Suzuki, T., Kuromori, M., Ichihara, H.: The Motion Capture System. IPSJ SIG Technical Reports Japan, CVIM128-13 (2001)
2. Hachimura, K., Nakamura, M.: Generation of Labanotation Dance Score from Motion-Captured Data. IPSJ SIG Technical Reports Japan, CVIM 128–14 (2001)
3. Arakaki, T., Hoshino, K.: Gneration of CG Dance Motion Synthesize Subjective Impressions. Technical Report of IEICE Japan, HIP2001-87 (2002)
4. Kamisato, S., Shiroma, T., Hoshino, K.: Estimation of Subjectivity in the Dancing Movement of Arm Motion and Generation of Dance Movements by Humanoid Robotic Arm. Technical Report of IEICE Japan, HIP2001-88 (2002)
5. Eva HIRES User's manual, Ver.5.0
6. AViUtl: http://www.ruriruri.zone.ne.jap/aviutl/
7. FAviAdjst: http://www.homepage1.nifty.con/nogue/

Discovering Areas of Expertise from Publication Data

Meredith Taylor and Debbie Richards

Computing Department, Maquarie University
North Ryde, Australia, 2109
richards@ics.mq.edu.au

Abstract. Expertise recommender systems are a valuable tool for keeping track of who has expertise and in what areas within an organization. The key problem is acquiring validated knowledge of expertise and keeping that information up to date. In research organizations, publications are one source of evidence of expertise which can be used to identify who knows about what. In this paper we focus on evaluating the feasibility of a simple technique for uncovering expertise used as the foundation and starting point of maintaining a profile of validated expertise within an organization.

Keywords: Expertise recommender system, knowledge acquisition, knowledge validation.

1 Introduction

Recommender systems are electronic (often web-based) systems that recommend objects of interest to searchers based on search queries (e.g. Amazon.com recommends books and other products, imdb.com recommends films). Once a searcher has selected a recommendation, the system will often recommend other things that it thinks the searcher will like based on the preferences of other searchers who also chose that recommendation. This is called Collaborative Filtering and works on the assumption that if person a likes one thing (or several things) that person b likes, it's likely that person a will also like other things that person b likes.

The term Recommender system was first introduced by Resnick and Varian [12] [11]; however, the first recommender system was Tapestry [5]. Tapestry was an electronic mail system which filtered the mail sent to searchers, returning only those messages that were of interest. Searchers specified how they wanted their mail filtered by providing search queries that the system ran over the set of documents. In addition to content filtering, Tapestry also used (and first coined the phrase) collaborative filtering. A searcher could pick out messages of interest, which could then be sent on to other searchers on the same mailing list who would have received those messages, but were not sure if they wanted to read them. Recommender systems can also be used to recommend experts. Expert Recommender Systems (ERS) are the focus of this project.

1.1 Expert Recommender Systems Issues

Expert recommender systems (systems that recommend experts) are a useful and convenient tool for finding experts [1] without having to spend time combing the

D. Richards and B.-H. Kang (Eds.): PKAW 2008, LNAI 5465, pp. 218–230, 2009.

internet, staff web pages, or publication repositories. As in any system that stores data, collecting the appropriate data can be hard and time consuming. Collecting the knowledge held in people's heads on who is an expert and in what area is harder still. Even more problematic is ensuring that the data obtained are complete, consistent and current so that the system can provide accurate recommendations. If a user is not certain that the recommendations provided by the system are valid, they will have little reason to trust the system, and thus the system will not be used.

To handle data and knowledge acquisition, two main approaches are used in expert recommender systems:

1.   Manual: Experts are required to register their own areas of expertise with the system by filling in surveys or entering keywords that can be matched with search queries; and

2.   Automatic: Data mining and other information retrieval techniques are used to search through sources that may hold evidence of peoples' expertise (such as web pages, publication repositories, citation indexes, and conference proceedings) to determine if someone is an expert in a certain field and to what extent they are an expert in that field (see for instance [2], [10], and [11]).

The first type of system is fairly easy to implement on a technical level as the experts themselves have to do most of the work by entering their own areas of expertise. The system developer will only need to develop appropriate forms for data entry and retrieval, and a simple query matching algorithm. Thus when a searcher searches the system, all that needs to be done is to match their search terms with the expert's keywords. This technique is often referred to as a yellow-pages approach to finding an expert as that is the way people usually find a plumber, lawyer or doctor. It is a simple and yet effective method for finding people who have certain skills. It uses the assumption that only someone who is actually an expert in a certain field would list themselves as such and since they went to the trouble of registering their areas of expertise, they are probably interested in being contacted. This assumption is not always valid. An expert may initially enter their data into the system, but may not perform any regular updates due to a lack of time or interest. As a result, people using the system can never be certain that a recommended expert's expertise is current, or that they are still willing to be contacted. It is even often the case that the expert will leave the organisation and their data will still be in the system. There are also the issues of experts failing to find the time to enter their data in the first place or entering incorrect data that does not reflect their true levels of expertise. However, this is a less likely occurrence than someone simply not entering their details initially since most members of an organisation would have a fairly realistic view of their level of expertise and would not wish to be contacted by someone if they are not confident that they would be able to help them.

The second type of system is less reliant on the time and interest of the expert. In some cases an automated system may provide a less biased profile of someone's expertise, but this depends on the appropriateness and range of sources available and information extraction techniques used. However, these systems are more difficult to implement as they require a large amount of data to be available for each expert. Additionally, expertise could be identified from many different sources that will vary across individuals and organisations making it difficult to have predefined sources for the system to search through. Sources that are used in existing systems to locate experts include email [3, 6], bulletin boards [7], web pages [4, 9], program code [8, 13], and technical reports [2].

To date we have developed a prototype system known as "Who Knows?" We have implemented and tested selected components of our proposed solution as initial proofs of concept. In the remainder of the paper we present our results to evaluate the feasibility of capturing initial data from artefacts and having experts validate the results. In the larger framework, our approach will encompass more sophisticated automated methods using a range of inputs such as individual web pages, project/grant repositories, citation indexes (e.g. CiteSeer - http://citeseer.ist.psu.edu/) and publications databases. In the longer term we intend to create a toolkit or workbench (like WEKA) which draws together the body of disparate work in this area by incorporating many algorithms and automated techniques such as those cited above. In the shorter term for proof of concept we have used a simple text analysis approach and an internal data source comprising a collection of all publications, grants and impact factors of individuals within our university.

2 Evaluating Automatic Expertise Acquisition

The Research office (RO) at Macquarie university runs and maintains IRIS -Integrated Research Information System (http://www.research.mq.edu.au/ researchers/iris) in which staff are required to enter information about all their publications from 2001 (the year the system was first put into use) onwards. The system also stores information about each staff member's research projects and grants (accepted and rejected) in the profile for the staff member. In their profiles, staff members are able to nominate RFCD (Research Fields, Courses and Disciplines) codes that correspond to their areas of expertise as well as the percentage of expertise they have in each area (Fig. 1).

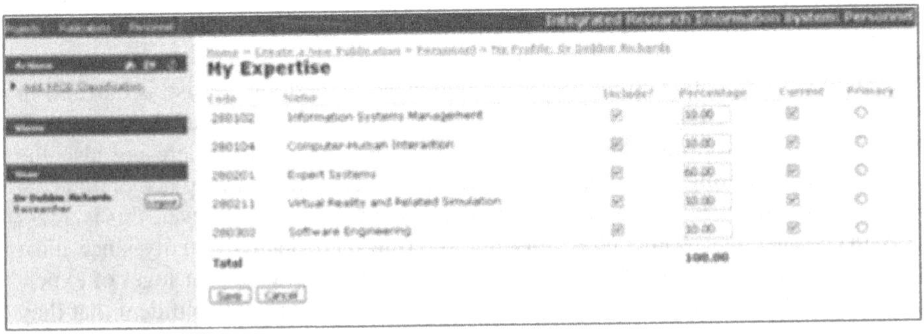

Fig. 1. Areas of expertise in IRIS

RFCD codes are issued by the Australian Research Council (ARC[1]) in order to categorise research and development activity and other activity within the higher education sector in a uniform manner. They are split into divisions (for example:

---

[1] see http://www.arc.gov.au/applicants/codes.htm

250000 - Chemical Sciences, 260000 - Earth Sciences, 420000 - Language and Culture, and 280000 - Information, Computing and Communication Sciences) which are then split into subdivision (such as 280101 - Information Systems Organisation and 280102 - Information Systems Management, which are subdivisions of Information, Computing and Communication Sciences).

Classifying experts with RFCD codes gives an indication of their general areas of expertise and would be a good addition to an expert's profile in an expert recommender system. Since few staff members have entered this data it is not possible to obtain this information directly from IRIS. If it was possible to automate the process of locating the RFCD codes for experts, it would not only provide a useful addition to each expert's profile in our prototype system, but it would also provide each expert with a more realistic view (that is an evidence-based view) of what their expertise areas actually are. Therefore we chose to use the publication data within IRIS to classify each publication with an RFCD code and then to assign RFCD codes to each staff member in IRIS based on the RFCD codes for each staff member's publications.

The publication information contained in IRIS includes details such as the name of the publication, the name of the publication it belongs to (in the case of journal and conference papers, for example), the author's name or list of authors' names, the primary department the publication belongs to, and the year of publication. It does not include paper abstracts, relevant keywords, or any online locations of papers. If we had this additional information we expect we could achieve better results and the effort of incorporating alternative and advanced algorithms would be more appropriate.

2.1 Methodology

The tasks involved in classifying experts with RFCD codes are the following:

1. Match RFCD codes with paper titles and publication titles using a simple string matching algorithm that checks to see if a keyword (or several keywords) in an RFCD code title occurs in the title of a publication or paper. (In this study this was done only for publications from the Computing Department).
2. Classify each staff member with the major RFCD codes found. (In this study we classified on the smaller division level, rather than the subdivision level or the major division level.
3. Check against self reported codes. It has been necessary to request assistance from members of the Computing Department for this exercise by asking them to classify their areas of interests with RFCD codes.
4. Record the percentage of experts that agreed with their automatically found codes.

Several Python scripts were written to complete these tasks which will now be described in more detail below.

2.2 Matching RFCD Codes with Paper and Publication Titles

This was done in several stages. The first stage was to collect the relevant data from the XML file that held the IRIS publication data. For the purposes of this experiment, only papers written by people from the Computing Department were considered. From these entries, the title of the paper (or book) was extracted, as well as the title of any accompanying publication (such as the journal or conference a paper was published in) and the list of authors.

The second stage was to collect the names of the staff currently in the Computing Department at Macquarie University. When this list of names was compiled, those staff members who did not have any publications in the list created from the IRIS data were eliminated. Similarly, publications in the list of IRIS data that were not authored by at least one person from the list of staff members were eliminated. The result of this process was a Python dictionary associating staff members with the publications that they had authored, co-authored, or edited.

The third stage involved gathering the relevant RFCD codes from the Australian Bureau of Statistics Website (http://www.abs.gov.au/). This collection was done prior to the new RFCD codes being released, thus the division used for matching was 280000 - Information, Computing and Communication Sciences. While some members of the Computing Department do have publications written in other domains, we felt that restricting the classification to one domain would simplify the process and show any significant results fairly quickly.

One goal of this experiment was to test what information from the IRIS publication data would provide the most accurate and predictable classifications. To this end, we classified each staff member's documents 3 times, once only using the paper (or book) titles, once using the containing publication titles (if applicable) and once using both paper and publication titles.

Matching RFCD codes to paper or publication titles was a fairly simple task. Each RFCD code was split into words. Then, each word was tested against the title in question using a simple string search. If the word was found, then that RFCD code was counted as a match. The only exception to this rule was the word 'computer' which is a common word to use in the domain and would have yielded too many false matches.

Minor tweaking of the string matching process was also performed to match words that share the same root (to make 280504 - Data Encryption match a paper with 'cryptography' in the title, for example). This was achieved by creating an ontology of terms found in the RFCD codes along with several words that share the same root and seem likely to appear in a publication title (Fig. 2).

```
"simulation": ["simulating", "simulate", "simulations"],
"analysis": ["analyse", "analysing"],
"representations": ["representing", "representative"],
"encryption": ["encrypting", "cryptography", "encoding",
"decryption",    "decoding", "cryptology"],
"security": ["secure", "unsecure", "secret"],
```

Fig. 2. Snippet of code from the ontology of terms, written as a dictionary in Python

The ontology also matched terms in the RFCD codes with words that referred to similar concepts. For instance the concept of a knowledge-based system is the same as for an expert system (RFCD 280201 Expert Systems). Thus the term 'expert' in the dictionary was matched with the term 'knowledge-based'. While this would not be a realistic task if we wished to classify staff from all disciplines, it was fairly simple to implement for only one discipline, and serves to show the possibility of such a task.

2.3 Classification

In the initial process of matching publication titles with RFCD codes, an attempt was made to match each staff member's publications with one or several RFCD codes. The codes used to match the publications were both subdivisional and divisional codes (e.g. 280100 Information Systems, and 280101 Information Systems Organisation). Thus each publication had on average three lists of codes associated with it: one list of codes matched purely on the title of the paper or book, one list of codes matched on the title of the containing publication, and one list matched on both titles. Because we were interested in classifying staff members' areas of expertise rather than their publications, we needed to gather the individual results together to provide a general classification of expertise for each staff member.

```
80 papers classified out of 95 by publication only
here are the rfcd codes and their count:
280500 Data Format: 30
280100 Information Systems: 182
280200 Artificial Intelligence and Signal & Image Processing: 53
280400 Computation Theory and Mathematics: 3
280300 Computer Software: 27

72 papers classified out of 95 by paper title
here are the rfcd codes and their count:
280500 Data Format: 49
280100 Information Systems: 132
280300 Computer Software: 13
280400 Computation Theory and Mathematics: 14
280200 Artificial Intelligence and Signal & Image Processing: 36

90 papers classified out of 95 by paper title and publication
here are the rfcd codes and their count:
280500 Data Format: 79
280100 Information Systems: 314
280300 Computer Software: 40
280400 Computation Theory and Mathematics: 17
280200 Artificial Intelligence and Signal & Image Processing: 89
```

Fig. 3. Example of RFCD classification for a staff member

Since this method of matching would match an RFCD code on only one word in the title, each publication could potentially yield many matches. We decided to simplify the output by classifying staff members' areas of expertise under the divisional (e.g. 280100 Information Systems) rather than the subdivisional codes (e.g. 280101 Information Systems Organisation). This involved adding up the number of subdivision matches under each division. This was not as straightforward as it seems. A paper with the word "information" in the title, for example, would match once with 280100 Information Systems, 280101 Information Systems Organisation, 280102 Information Systems Management, 280103 Information Storage, Retrieval and Management, and 280112 Information Systems Development Methodologies. Since each of these RFCD codes was counted as a match, the final count for the major division 280100 Information Systems would be 5 for this paper. However, giving such a large weighting to an RFCD code based only on one word would be misleading. Thus, since the same one word was matched from each of these codes, the final count for the major division 280100 Information Systems should be 1 for this paper (if the words 'information' and 'systems' were present, the count would be 2). Thus the number of matches associated with each division was altered to reflect the proportion of words in the title of the paper or publication that yielded the match.

The final output for the classification process is three sets of classifications for each expert: one set of classifications showing the divisional RFCD codes that were matched on the paper titles along with their counts, one showing the codes that were matched on the containing publication titles, and one showing the codes that were matched on both titles (Fig. 3).

2.4 Validating the Results

After automatically classifying each staff member's areas of expertise, we then needed to have staff members view the classifications and accept or reject them. We selected 20 staff members each with more than 10 publications and sent them their results asking them to indicate which they felt was correct and incorrect. We also asked them to select RFCD codes from a list provided to them that they felt most accurately represented their areas of research.

Table 1. Statistics for total number of publications for each staff member

	No. Publications
Mean	15.80702
Stdev	27.41509
Median	6
Mode	1
Max	169

On the 31st of March 2008, a new set of RFCD codes were released to be used from April 1 onwards. When we discovered this, the emails with the old RFCD codes had already been sent to all 20 staff members, and 10 had replied. We decided to classify the remaining 10 staff members with the new RFCD codes and resend them their results. The nature of the algorithm we used was such that it could just as easily classify staff publications using the new codes as it did with the old ones. As the introduction of the new codes was fairly recent, many systems and institutions are still using the old RFCD codes,[2] so we felt that classifying under these codes is still relevant, but only for the short term. Additionally, classifying under both the old and new codes may give us good information about which set of codes more accurately classified the staff members with the algorithm we used.

---

[2] See, for instance the ARC website: http://www.arc.gov.au/applicants/codes.htm (last accessed: 9/6/08), and *Find an Expert* at the University of Melbourne: http://www.findanexpert.unimelb.edu.au/ (last accessed: 9/6/08)).

2.5 Results

There were 57 members of the Computing Department who had publication data listed in IRIS. Information about the total number of papers for each staff member is shown in Table 1. Fig. 4 shows the percentage of papers that were able to be classified using only the paper title, only the publication title, and both the publication and paper titles. The Expert IDs are sorted by their total number of publications from smallest to largest. It can be observed that the paper title only method in most cases consistently performed worse than both the publication title only method and the publication and paper title method, while the paper and publication title method consistently performed the best.

On average, the system was able to classify 96.15% of each staff member's papers with the new RFCD codes, and 96.04% with the old RFCD codes using both the paper and publication titles. Additional information about the number of publications classified for the old and new codes is shown in tables 2 and 3, respectively.

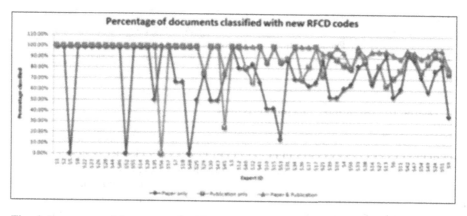

Fig. 4. Percentage of documents classified for each staff member via the three classification methods

Table 2. Percentage of documents classified with Old RFCD codes by classification method

Classification method	Mean	Stdev	Median	Mode	Max	Min
Paper title	70.55%	26.47%	71.79%	100.00%	100.00%	0.00%
Publication title	88.26%	18.35%	95.45%	100.00%	100.00%	0.00%
Paper & pub. title	96.04%	6.59%	100.00%	100.00%	100.00%	75.00%

Table 3. Percentage of documents classified with New RFCD codes by classification method

Classification method	Mean	Stdev	Median	Mode	Max	Min
Paper title	71.96%	26.65%	75.00%	100.00%	100.00%	0.00%
Publication title	88.26%	18.35%	95.45%	100.00%	100.00%	0.00%
Paper & pub. title	96.15%	6.48%	100.00%	100.00%	100.00%	75.00%

As can be observed in Tables 2 and 3, classifying on paper and publication title classified more documents on average than classifying on publication title only, which in turn classified more documents that classifying on paper title only. Table 4 shows the results of a Wilcoxon signed-rank test comparing each of the methods.[3] From this table it can be observed that that the three methods differ significantly in the average number of documents that they are able to classify. This indicates that, at least for publications authored by staff members in the Computing Department, more information about a publication than the title of the paper or book is needed. This is not surprising, as often a certain amount of creative license is taken with the title of a paper or book so that it may not be easily associated with the domain (e.g. "Training for High Risk Situations"). Conferences and journals, on the other hand, will generally contain domain specific keywords in the title (e.g. "Proceedings of Fourth International Joint Conference on Autonomous Agents and Multi Agent Systems").

Table 4. Wilcoxon signed rank test comparing percentage of documents classified by method x against percentage of documents classified by method

Method of classification 1 (x)	Method of classification 2 (y)	W	ns/r	P(1-tail)	P(2-tail)	z
Paper title	Publication title	-481	36	0.0001	0.0002	-3.77
Paper title	Paper & Pub. titles	-780	39	<.0001	<.0001	-5.44
Publication title	Paper & Pub. titles	-276	23	<.0001	<.0001	-4.19

2.6 Testing the Dictionary of Similar Words

We also wanted to test if our dictionary of words in the RFCD codes and words that share the same root, or refer to similar concepts, was able to classify more documents than if we hadn't used it. Fig. 5 shows the percentage of papers classified by paper title and publication with the old codes using the similar word dictionary versus the percentage classified without the similar word dictionary with the experts sorted on total number of papers. We can see that in most cases using the similar word dictionary will classify more documents than not using it, and never less. In fact, using the similar word dictionary classified on average 32.24% more papers with the old RFCD codes than not using it in our algorithm.

Fig. 6 shows the percentage of papers classified by paper title and publication with the new RFCD codes using the similar word dictionary versus the percentage classified without the similar word dictionary with the experts again sorted on total number of papers. We can see that in most cases the two methods classified an equal or very similar number of documents, even when the number of documents got quite large. In fact, using the similar word dictionary will classify on average only 2.83% more papers with the new RFCD codes than not using it with our algorithm.

This indicates that the new RFCD codes (at least in the division that we used) are more suited to our classification task than the old with regards to the number of classifications made. It is also pleasing that the changes by the ARC to the codes are more

---

[3] The Wilcoxon signed-rank test is a non-parametric alternative to a paired t-test. This type of test was used instead of a t-test, as the population could not be assumed to be normally distributed.

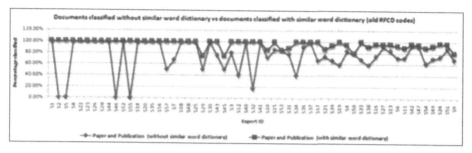

Fig.5. Percentage of documents classified with old RFCD codes for each staff member with and without the similar word dictionary for the paper and publication title method of classification

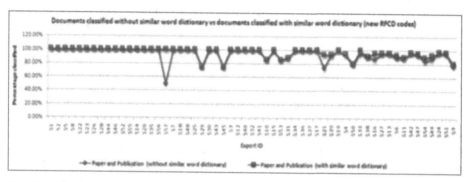

Fig. 6. Percentage of documents classified for each staff member with and without the similar word dictionary for the paper and publication title method of classification (new RFCD codes)

reflective of current research activity. Also very relevant is that there are 102 new RFCD codes as compared to 46 old RFCD classification codes.

2.7 Comparing Classifications against Staff Members' Responses

18 out of 20 of the staff members that we contacted responded (90% response rate): 10 with the old RFCD codes and 8 with the new RFCD codes. We asked the experts to indicate which of the RFCD codes they were classified with they felt were correct and which they felt were incorrect. We added the counts of each correct RFCD code together for each classification method (paper title, publication title, paper and publication title) to get an accuracy score for each expert. We then were able to calculate the percentage of accuracy between our classification and what the expert felt was correct.

From analysis of the individual data for each expert we found that all experts except one agreed with their highest ranked RFCD code. The expert who disagreed, S33, said that some of their work could not be classified under the RFCD division 280000 - Information, Computing and Communication Sciences, but rather under 410300 Cinema, Electronic Arts and Multimedia due to their work in computer games. However, they only had one document listed in IRIS that came under the category of computer games (i.e. that had anything to do with computer games in the title). It may be possible that this staff member had written papers on computer games before 2001, and thus these papers were not entered into IRIS, or after 2006, in which case we would not have had access to these papers as we were not provided with any data about publications written after this period. This does, however, raise the issue of the extent to which the system should accept an expert's validation of the data produced by automated searching. If the system simply accepts anything the expert says, despite there being no available evidence for it, it may very well experience some of the same problems faced by systems that rely entirely on self-reporting by experts. However, if the system refuses to accept any changes to the automated searching results unless it is sure that there is evidence somewhere that validates these changes, it may reject experts whose work is either too new, too old, or in a medium or location that the system does not search. While it is clear that a compromise is needed, it is not yet certain how this compromise can be reached or even if such a compromise is possible.

While 17 out of the 18 experts (94.4%) agreed with their highest ranked RFCD code, only 7 out of the 18 (38.9%) agreed with their second highest ranked code. The average percentage of accuracy of the system's codes was 62.05% overall, and 65.86% for the old codes and 57.30% for the new codes (see Table 5). The larger average for the old codes can probably be attributed to two factors: firstly, a couple of staff members who were sent the old RFCD codes responded very generally that they agreed with everything or thought that the classifications were 'good enough'. Secondly, a greater number of codes were assigned to each staff member when using the new codes. This is because the new RFCD codes not only have more categories, but some RFCD codes previously in the division 280000 - Information, Computing and Communication Sciences were moved to other divisions, so it was necessary to include these in the classification algorithm as well. Since the old RFCD codes did not have that many categories under division 280000, the staff members did not have as many codes to choose from when classifying themselves.

Table 5. Percentage of accuracy by RFCD code type

Code Type	Mean	Stdev	Min	Max
Old	65.86%	26.51%	24.42%	100.00%
New	57.30%	22.29%	18.75%	88.98%
Both	62.05%	24.41%	18.75%	100.00%

The fairly low averages for both the old and new RFCD codes could also be attributed to the staff members not having a clear understanding of what the codes meant. As each staff member undoubtedly had a preconceived notion of what their areas of research or expertise were, they may not have considered what each code actually represented, and whether a paper they authored may actually have fallen under a different RFCD code than they expected. On staff member, S51 mentioned that they had thought that most of their work could be classified under 280200 Artificial Intelligence and Signal and Image Processing, but seeing that the system had found their highest ranked code to be 280100 Information Systems, they realised that a lot of their recent work could be classified under this code

In general, this algorithm appears to be quite successful in determining the most prominent RFCD code for staff members in the Computing Department at Macquarie

University. It is not certain how this algorithm would perform for staff members in other departments or at other universities. Further work would need to be done to refine the algorithm to increase its accuracy and to test the algorithm on publications from other departments. However, there the accuracy of this and any other algorithm is limited by only having the titles of the paper and publication as input. Ideally the abstract of the document, or even the document itself should be included for the most accurate results.

3 Summary and Conclusions

This paper has described and implemented a simple method for automatic identification and population of a repository of experts and areas of expertise. It has described an experiment where members of the Computing Department were assigned several RFCD codes relating to their areas of expertise using their publication titles and a fairly simple string-matching algorithm. In summary, the algorithm was quite successful at determining each staff member's most prominent RFCD code, although in most cases it did make some false predictions. After each staff member in the Computing Department who had any documents associated with their names in IRIS had been classified with RFCD codes, 20 staff members with more than 10 documents each were sent their results. Each staff member was asked to indicate if the classifications were accurate and, if not, which RFCD codes they would use to classify their areas of research. A large percentage of staff members responded (90%) and all but one agreed with the RFCD code that was considered most relevant by the system to their areas of expertise. This was considered to be a very promising result from a fairly simple and crude method which, if developed further, could produce even more promising results. Our approach offers a viable alternative to relying on experts to enter and maintain their own data with the outcome being a validated set of recommendations based on integrated machine and human effort.

Acknowledgements. Thanks to the Computing Department staff for their participation in this project and the numerous others who participated in usability studies and interviews.

References

[1] Aïmeur, E., Onana, F., Saleman, A.: HELP: A Recommender System to Locate Expertise In Organizational Memories. In: 2007 ACS/IEEE International Conference on Computer Systems and Applications, AICCSA 2007, Amman, Jordan, pp. 866–874 (2007)

[2] Crowder, R., Hughes, G., Hall, W.: An agent based approach to finding expertise. In: Proceedings of the Fourth International Conference on Practical Aspects of Knowledge Management, Heidelberg, Germany, pp. 179–188 (2002)

[3] Ehrlich, K., Lin, C.-Y., Griffiths-Fisher, V.: Searching for experts in the enterprise: combining text and social network analysis. In: GROUP 2007: Proceedings of the 2007 international ACM conference on Supporting group work, pp. 117–126. ACM, New York (2007)

[4] Foner, L.: Yenta: A Multi-Agent Referral-Based Matchmaking System. In: AGENTS 1997: Proceedings of the 1st International Conference on Autonomous Agents Marina del Rey California, pp. 301–307 (1997)

[5] Goldberg, D., Nichols, D., Oki, B.M., Terry, D.: Using collaborative filtering to weave an information tapestry. Communications of ACM 35(12), 61–70 (1992)

[6] Kautz, H., Selman, B., Shah, M.: Referral Web: Combining Social Networks and Collaborative Filtering. Communications of ACM 40(3), 63–65 (1997)

[7] Krulwich, B., Burkey, C.: The ContactFinder Agent: Answering Bulletin Board Questions with Referrals. In: Proceedings of the 1996 National Conference on Artificial Intelligence (AAAI 1996), Portland, Oregon, vol. 1, pp. 10–15 (1996)

[8] McDonald, D.W., Ackerman, M.S.: Expertise recommender: a flexible recommendation system and architecture. In: Proceedings of the 2000 ACM Conference on Computer Supported Cooperative Work (CSCW 2000), Philadelphia, Pennsylvania, United States, December 2-5, pp. 231–240 (2000)

[9] Pikrakis, A., Bitsikas, T., Sfakianakis, S., Hatzopoulos, M., DeRoure, D., Reich, S., Hill, G., Stairmand, M.: MEMOIR – Software Agents for Finding Similar Users by Trails. In: Proceedings of the 3rd International Conference on Practical Application of Intelligent Agents and Multi-agents, London, UK, pp. 453–466 (1998)

[10] Prekop, P.: Supporting Knowledge and Expertise Finding within Australia's Defence Science and Technology Organisation. In: Proceedings of the 40th Annual Hawaii International Conference on System Science (HICSS-40), Hawaii, January 3-6 (2007)

[11] Reichling, T., Schubert, K., Wulf, V.: Matching human actors based on their texts: design and evaluation of an instance of the ExpertFinding framework. In: Proceedings of the 2005 International ACM SIGGROUP Conference on Supporting Group Work, Sanibel Island, Florida, USA, November 6 - 9, pp. 61–70 (2005)

[12] Resnick, P., Varian, H.R.: Recommender systems. Communications of ACM, special issue 40, 56–58 (1997)

[13] Vivacqua, A.: Agents for Expertise Location. In: Proceedings of AAAI Spring Symposium on Intelligent Agents in Cyberspace, Stanford, CA, pp. 9–13 (1999)

Facial Feature Extraction Using Geometric Feature and Independent Component Analysis

Toan Thanh Do[1] and Thai Hoang Le[2]

[1] Department of Computer Sciences, University of Natural Sciences, HCMC, Vietnam
dttoan@fit.hcmuns.edu.vn
[2] Department of Computer Sciences, University of Natural Sciences, HCMC, Vietnam
lhthai@fit.hcmuns.edu.vn

Abstract. Automatic facial feature extraction is one of the most important and attempted problems in computer vision. It is a necessary step in face recognition, facial image compression. There are many methods have been proposed in the literature for the facial feature extraction task. However, all of them have still disadvantage such as not complete reflection about face structure, face texture. Therefore, a combination of different feature extraction methods which can integrate the complementary information should lead to improve the efficiency of feature extraction stage. In this paper we describe a methodology for improving the efficiency of feature extraction stage based on the association of two methods: geometric feature based method and Independent Component Analysis (ICA) method. Comparison of two methods of facial feature extraction: geometric feature based method combined with PCA method (called GPCA) versus geometric feature based method combined with ICA method (called GICA) on CalTech dataset has demonstrated the efficiency of GICA method. Our results show that GICA achieved good performance 96.57% compared to 94.70% of GPCA method. Furthermore, we compare two methods mentioned above on our dataset, with performance of GICA being 98.94% better 96.78% of GPCA method. The experiment results have confirmed the benefits of the association geometric feature based method and ICA method in facial feature extraction.

Keywords: Face recognition; independent component analysis (ICA); principal component analysis (PCA); geometric features.

1 Introduction

Face recognition has a variety of potential applications in public security, law enforcement and commerce such as identity authentication for credit card or driver license, access control, information security, and video surveillance, etc.. In addition, there are many emerging fields that can benefit from face recognition such as human-computer interfaces and e-services, including e-home, tele-shopping and tele-banking.

However, one of the most difficult from face recognition problem is the facial feature extraction step. A good feature extraction will increase performance of face recognition system. Various techniques have been proposed in the literature for this purpose, and are mainly classified in four groups.

D. Richards and B.-H. Kang (Eds.): PKAW 2008, LNAI 5465, pp. 231–241, 2009.
© Springer-Verlag Berlin Heidelberg 2009

1.1 Geometric Feature Based Method Group

The features are extracted by using relative positions and sizes of the important components face such as eyes, nose, mouth and other important component face. Almost this group concentrates in two directions.

Firstly, detecting edges, directions of important components or region images contain important components, then building feature vectors from these edges and directions. Using filters such as Canny filter to detect eyes or/and mouth image region, using transforms method such as Hough transform to detect eyes [1] and the gradient analysis is a method which is usually applied in this direction.

Secondly, methods are based on the grayscales difference of important components and unimportant components, by using feature blocks, set of Haar-like feature block [2] in Adaboost method to change the grayscales distribution into the feature vector. One of the proposed methods is LBP method [4]; it divides up the face image to regions (blocks); each region corresponds with each central pixel; we examine its pixel neighbors, based on the grayscales value of central pixel to change its neighbor to 0 or 1. Therefore, every pixel will be represented in a binary string. Since then, we build histograms for every region. Then these histograms are combined to a feature vector for the face image. The two methods, Cascaded AdaBoost [3] and Chain AdaBoost, use a set of Haar-like features to detect important components of face or/and face. Selecting features is based on different grayscales or "integral image" in every Haar-like feature block. In Gabor Wavelets method [5], Gabor filter can capture salient visual properties such as spatial localization, orientation selectivity, and spatial frequency characteristics. The advantages of these methods are concentration on important components of face such as eyes, nose, mouth, etc. but the disadvantage is not to remain face global structure and face texture.

1.2 Template Based Method Group

Based on a template function and appropriate energy function, this method group will extract feature of important component of face such as eyes, mouth, etc. or face shape.

An image region is the best appropriateness with template (eyes, mouth, etc.) which will make minimize the energy. The methods have been proposed such as deformable template and genetic algorithms. In the deformable template method [6] [7] [8], the feature of interest, an eye, for example, is described by a parameterized template. An energy function is defined to links edges, peaks, and valleys in the image intensity with corresponding properties of the template. Then the template interacts dynamically with the image, by altering its parameter values to minimize the energy function, thereby deforming itself to find the best fit. The final parameter values can be used as descriptors for the features. Advantages of this group method are using template and determining parameter for important components of face, but disadvantage is not to reflect face global structure.

1.3 Color Segmentation Based Method Group

This group method based on skin's color to isolate the face .Any non-skin color region within the face image region is viewed as a candidate for "eyes" and/or "mouth" [9] [10].

1.4 Appearance Based Method Group

Goal of this method group is using linear transformation and statistical methods to find the basic vectors to represent the face. Methods have been proposed in the literature for this aim such as PCA [12] and ICA [13] [15].

In detail, goal of PCA method is to reduce the number of dimensions of feature space, but still to keep principle features to minimize loss of information. PCA method uses second-order statistic (covariance matrix) in the data. However, PCA method has still disadvantages. High order dependencies still exist in PCA analysis, for example, in tasks as face recognition, much of the important information may be contained in the high-order relationships among the image pixels, not only second-order. Therefore, we need to find a method more general than PCA and ICA [14] is a satisfying method. Instead of principle component analysis, ICA uses technique independent component analysis – an analysis technique not only use second-order statistic but also use high order statistic (kurtosis). PCA can be derived as a special case of ICA which uses Gaussian source models. In this case, the mixing matrix cannot determine. PCA is not the good method in cases non Gaussian source models. In particular, it has been empirically observed that many natural signals, including speech, natural images, are better described as linear combinations of sources with "super-Gaussian" distributions (kurtosis positive). In this case, ICA method better than PCA method because: 1) ICA provides a better probabilistic model of the data. 2) It uniquely identifies the mixing matrix. 3) It finds an unnecessary orthogonal basic which may reconstruct the data better than PCA in the presence of noise such as variations lighting and expressions of face. 4) It is sensitive to high-order statistics in the data, not just the covariance matrix. The appearance based method group has been found the best performer in facial feature extraction problem because it keeps the important information of face image, rejects redundant information, reflects face global structure.

In many application of face recognition such as identity authentication for credit card and video surveillance, accuracy of face recognition problem is the best important factor. Therefore, beside principle components, independent components of data are kept by PCA and ICA method. Combination of these features with geometric features such as nose, eyes and mouth in face recognition will make increase accuracy, confident of face recognition system because they complete reflection either face structure or face texture. In this paper, we present combining ICA method with geometric feature based method, then comparison of GICA method and GPCA method on CalTech dataset and our dataset. Since then, practicability of GICA method for face recognition problem is demonstrated.

The remainder of this paper is organized as follows. In section 2, we present about combination of face important components features. In section 3, we present architecture II to apply ICA method for face extraction features. In section 4, we present our method for combination of geometric feature based method and ICA method in facial feature extraction. Experimental results are reported in section 5. Conclusion and future works will be mentioned in section 6.

2   Combinations of Facial Important Components Features

2.1   Face Global Feature Vector

Firstly, we normalize the face images in a standard size of 30x30 pixels (figure 1).

Fig. 1. Face global standardization

After normalization the face images will be represented by vectors $x_{face} = (x_1, x_2,...,$ $x_{900})^T$. Then, vectors x_{face} will be mapped into a face feature space. Vectors representing faces in the face feature space is $y_{face} = (y_1, y_2,...,y_k)^T$, where k is dimension in the face feature space; y_{face} is called "global feature vector".

2.2   Face Component Feature Vectors

In [11], the authors presented a method to detect image regions that contain eyes, mouth, in face color images. To detect image regions containing left eye, right eye, mouth, firstly, they have used filter technique and threshold to change images to binary images, then using connected region algorithm to detect image regions which contain eyes, mouth. Figure 2 shows an example about detection eyes, mouth from face images.

Fig. 2. Extraction face important components

After locating image regions of eyes, mouth in face image, we have vectors that represent left eye image (28x30), right eye image (28x30), mouth image (20x40): $x_{left\_eye} = (x_1, x_2,...,x_{840})^T$, $x_{right\_eye} = (x_1, x_2,...,x_{840})^T$, $x_{mouth} = (x_1, x_2,...,x_{800})^T$. Then, mapping face component feature vectors to component feature space, we have $y_{left\_eye} = (y_1, y_2,..., y_k)^T$, $y_{right\_eye} = (y_1, y_2,..., y_k)^T$, $y_{mouth} = (y_1, y_2,..., y_k)^T$. These vectors are called "component feature vector"

2.3   Combination of Face Global Feature Vector and Face Component Feature Vector

By combining "global feature vector" and "component feature vector", we get "combination vector" as follows:

y_{comb} will be featuring vector that used for face recognition step.

$$y_{comb} = \begin{bmatrix} y_{face} \\ y_{left\_eye} \\ y_{right\_eye} \\ y_{mouth} \end{bmatrix}$$

Fig. 3. Combination face global feature and face important components feature

3   ICA Method for Facial Feature Extraction

Independent Component Analysis (ICA) minimizes both second-order and higher-order dependencies in the input data and attempts to find the basis along which the data (when projected onto them) are - statistically independent. Bartlett et al [15] provided two architectures of ICA for face recognition task: Architecture I - statistically independent basis images, and Architecture II - factorial code representation.

In a face recognition problem, our goal is to find coefficients of feature vectors to achieve the most independent in desire. Therefore, in this paper, we selected architecture II of ICA method for the face representation. A number of algorithms for performing ICA have been proposed (see [16] for reviews). In this paper, we apply FastICA algorithm developed by Aapo Hyvärinen [16] for our experiments.

Architecture II: Statistically Independent Coefficients. The goal in this approach is to find a set of statistically independent coefficients.

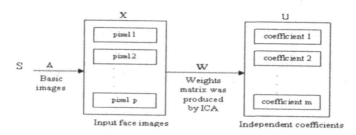

Fig. 4. Finding coefficients which presentation images are independent

We organize the data matrix X so that the images are in columns and the pixels are in rows. Pixel i and j are independent if when moving across the entire set of images, it is not possible to predict the value taken by pixel i based on the corresponding value taken by pixel j on the same image. The goal in architecture I is using ICA to find a set of statistically independent basic images. Although basic images found in architecture I are approximately independent, when projecting down statistically independent basic images subspace, feature vectors of each image are not necessarily independent. Architecture II use ICA to find a representation which coefficients are used to represent an image in the basic images subspace being statistically independent. Each row of weight matrix W is an image. A, an inverse matrix of W, contains basic images in its columns. Statistically independent coefficients in S will be recovered in columns of U (figure 4); each column of U contains coefficients for combination of basic images in A to construct images of X.

Architecture II is implemented through the following steps:

Assumption that we have n images; each image has p pixels. Therefore, data matrix X has an order of p × n.

1. Let R be a p × m matrix containing the first m eigenvectors of a set of n face images in its columns.

2. Calculating set of principle component of set of images in X:

$$C = R^T \times X \tag{1}$$

3. The coefficients for linearly combining the basic images in A are determined:

$$U = W \times C \tag{2}$$

Assumption that we have a set of images for testing X_{test}, feature extraction of X_{test} is computed through the following steps: firstly, from X_{test}, we calculate a set of principle component of X_{test} by:

$$C_{test} = R^T \times X_{test} \tag{3}$$

Then, a set of feature vectors of X_{test} in the basic images space is calculated by:

$$U_{test} = W \times C_{test} \tag{4}$$

Each column of U_{test} is a feature vector corresponding with each image of X_{test}.

Firstly, to face representation with ICA method, we apply PCA to project the data into a m dimensional subspace with purpose to control the number of independent components made by ICA, and then ICA is applied to the eigenvectors to minimize the statistical dependence of feature vectors in the basic images space. Thus, PCA uncorrelated input data, high-order dependence remain will be separated by ICA.

The first row in figure 5 contains the eight eigenvectors (eight eigenfaces) corresponding with eight the highest eigenvalues of PCA method and the second row includes the first of eight basic images in architecture II of ICA on CalTech dataset.

Fig. 5. First row contains eight eigenfaces corresponding eight the highest eigenvalues for PCA. Second row first of eight basic images for architecture II ICA.

4   The Association of Geometric Feature Based Method and ICA in Facial Feature Extraction

After detect image regions that contain eyes, mouth, in face color images. ICA method was applied to them and global face. Therefore, we get vectors which representation for eyes, mouth and global face in ICA subspace. The combination of these vectors

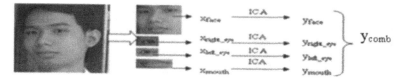

Fig. 6. Facial feature extraction by geometric feature based method combination with ICA method

will create vector which reflect texture and geometric feature of face image. The process of feature extraction in our study was illustrated in figure 6.

y_{comb} is a feature vector of combining the geometric feature based method and the ICA method. It will be used for face recognition step.

5   Experimental Results and Discussion

5.1   Datasets for Experiments

In this section we describe our experiments on two face datasets: the CalTech dataset and our self datasets.

CalTech dataset: The dataset includes Markus Weber's 450 color images at California Institute of Technology. After using AdaBoost to detect face regions, we get 441 face images of 26 people. We used 441 these images to perform GPCA and GICA. In CalTech dataset, almost images are frontal faces and have not the same illumination. Some images are too dark or too bright. Some images are hidden important components such as eyes. Some images have different face expressions. We divide up data set to two parts. Training set contains 120 images and testing set contains 321 images. We repeated our experiments for ten random divides up of the data set so that every image of the subject can be used for testing. Results were reported on the average performance. CalTech dataset is publicly available for research aims at the URL: http://www.vision.caltech.edu/html-files/archive.html. Figure 7 shows samples image from CalTech dataset

Fig. 7. Samples face images from CalTech dataset

Our dataset: Our dataset was built from two sources:

The students of Nguyen Huu Cau high school (Hoc Mon district, Ho Chi Minh city) and students of the faculty of information technology of HCMC University of Natural Sciences. There are 159 images of 19 people. In our dataset, most images are

Fig. 8. Samples face images from our dataset

left rotated ($\leq 30^0$), right rotated ($\leq 30^0$). Some of images have not the same illumination. We divide up data set to two parts. Training set contains 83 images and testing set contains 76 images. We repeated our experiments for ten random divides up of the data set so that every image of the subject can be used for testing. Results were reported on the average performance. Figure 8 shows samples image from our dataset.

The recognizer was implemented by the neural network method. Fast Artificial Neural Network is used in our experiment. Fast Artificial Neural Network Library (FANN), which is a free open source neural network library, implements multilayer artificial neural networks in C language and supports for both fully connected and sparsely connected networks. FANN has been used in many studies. FANN implementation includes:

Training step: assumption that we have n classes (n different people), training with FANN will create n sets of weights. Each set of weights corresponds with each class (each person).

Testing step: the input is a person's face image (one of the n people mentioned above); this face image was tested with n sets of weights which were created in the training step, this person belongs to a class which corresponding with the set of weights make the biggest output. FANN is publicly available for research aims at the URL: http://leenissen.dk/fann/

5.2 Results on the CalTech Dataset

Table 1 reports the results on CalTech dataset of two different algorithms applied to the face recognition. The recognition rate for GICA method shows a better recognition rate.

Table 1. Percentage accuracy values on the CalTech dataset.

Method	GPCA	GICA
Recognition rate	94.70	**96.57**

The average number of principal component for PCA representation of global face has 100; and average equal to 60 for left eye, right eye and mouth. The average number of independent component for ICA representation of global face has 100; and average equal to 60 for left eye, right eye and mouth.

The cumulative match score vs. rank curve is used to show the performance of each method, (figure 9). Here, cumulative match score and rank are the percentage accuracy that can be achieved by considering the first k biggest output (corresponds with k classes in dataset) of FANN. The rank is a reliability measure and it is very

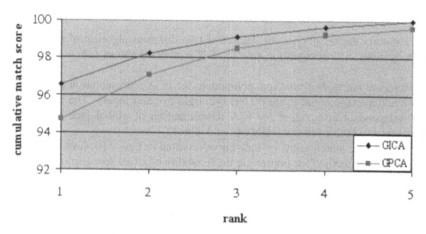

Fig. 9. Rank-curves on the CalTech dataset. Reported results show that the GICA method produces more reliable system.

Table 2. Percentage accuracy values on our dataset

Method	GPCA	GICA
Recognition rate	96.78	**98.94**

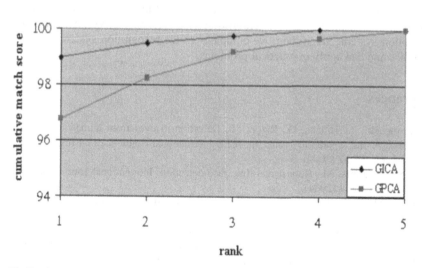

Fig. 10. Rank-curves on our dataset. Reported results show that the GICA method produces more reliable system.

important for video-surveillance applications in uncontrolled environments. Even in this case, the GICA method gives a sharp improvement of the performance in comparison with GPCA method.

5.3 Results on the Our Dataset

Table 2 reports the results on our dataset of two different algorithms applied to the face recognition. The recognition rate for GICA method shows a better recognition rate.

The average number of principal component for PCA representation of global face has 100; and average equal to for 60 left eye, right eye and mouth. The average number of independent component for ICA representation of global face has 100; and average equal to 60 for left eye, right eye and mouth.

The cumulative match score vs. rank curve reported in figure 10 show the effectiveness of the GICA method for improving the reliability of a face recognition system.

6 Conclusions

In this paper, a technique for automatic facial feature extraction based on the geometric features of human face and ICA method is presented.

With applying neural network method in the recognition step, this paper do comparisons, evaluations about GPCA and GICA method on CalTech dataset (containing 450 images) and our dataset (containing 159 images). Results from two methods facial feature extraction GPCA and GICA method on CalTech dataset are illustrated on table I and on our dataset are showed on table II. Through experiment, we notice that GICA method will have good results for cases of face input with the suitable brightness and position mentioned in section 5. In future, we continue experiment with other face dataset so that it demonstrates practicability of GICA method in face recognition system. Since then, we complete a face recognition system with general input face and can apply in practical problems.

References

1. Kawaguchi, T., Hidaka, D., Rizon, M.: Detection of eyes from human faces by Hough transform and separability filter. In: IEEE International Conference on Image Processing, vol. 1, pp. 49–52 (2000)
2. Viola, P., Jones, M.: Robust real-time face detection. International Journal of Computer Vision, 137–154 (2004)
3. Jones, M., Viola, P.: Face Recognition Using Boosted Local Features. In: IEEE International Conference on Computer Vision (2003)
4. Liao, S., Fan, W., Chung, A.C.S., Yeung, D.-Y.: Facial Expression Recognition Using Advanced Local Binary Patterns, Tsallis Entropies And Global Appearance Features. In: IEEE International Conference on Image Processing, pp. 665–668 (2006)
5. Liu, C., Wechsler, H.: Gabor Feature Based Classification Using the Enhanced Fisher Linear Discriminant Model for Face Recognition. IEEE Trans. Image Processing 11(4), 467–476 (2002)
6. Yuille, A.L., Cohen, D.S., Hallinan, P.W.: Feature Extraction From Faces Using Deformable Templates. In: IEEE Computer Society Conference on Computer Vision and Pattern Recognition, pp. 104–109 (1989)

7. Zhang, L.: Estimation Of The Mouth Features Using Deformable Templates. In: IEEE International Conference on Image Processing, vol. 3, pp. 328–331 (October 1997)
8. Kuo, P., Hannah, J.: An Improved Eye Feature Extraction Algorithm Based On Deformable Templates. In: IEEE International Conference on Image Processing, vol. 2, pp. 1206–1209 (2005)
9. Phung, S.L., Bouzerdoum, A., Chai, D.: Skin Segmentation Using Color And Edge Information. Signal Processing and Its Applications 1, 525–528 (2003)
10. Sawangsri, T., Patanavijit, V., Jitapunkul, S.: Face Segmentation Using Novel Skin-Color Map And Morphological Technique. In: Proceedings of World Academy of Science, Engineering and Technology, vol. 2 (January 2005) ISSN 1307-6884
11. Le, T.H., Nguyen, T.M., Nguyen, H.T.: Proposal of a new method of feature extraction for face recognition. National Conference about Information Technology, DaLat city (2006) (in VietNamese)
12. Turk, M., Pentland, A.: Face recognition using eigenfaces. In: IEEE Conference on Computer Vision and Pattern Recognition, pp. 586–591 (1991)
13. Draper, B.A., Baek, K., Bartlett, M.S., BeveRidge, J.R.: Recognizing Face with PCA and ICA. Computer Vision and Image Understanding 91, 115–137 (2003)
14. Comon, P.: Independent component analysis—A new concept? Signal Processing 36, 287–314 (1994)
15. Bartlett, M.S., Movellan, J.R., Sejnowski, T.J.: Face Recognition by Independent Component Analysis. IEEE Transactions on Neural Networks 13(6) (November 2002)
16. Hyvärinen, A., Oja, E.: Independent Component Analysis: Algorithms and Applications. Neural Networks, 411–430 (2000)

Author Index

Printed in the United States
By Bookmasters